Bo

W9-ASE-604

# ELECTRIFY

# ELECTRIFY

AN OPTIMIST'S PLAYBOOK FOR OUR CLEAN
ENERGY FUTURE

SAUL GRIFFITH

THE MIT PRESS   CAMBRIDGE, MASSACHUSETTS   LONDON, ENGLAND

© 2021 Massachusetts Institute of Technology

All rights reserved. No part of this book may be reproduced in any form by any electronic or mechanical means (including photocopying, recording, or information storage and retrieval) without permission in writing from the publisher.

The MIT Press would like to thank the anonymous peer reviewers who provided comments on drafts of this book. The generous work of academic experts is essential for establishing the authority and quality of our publications. We acknowledge with gratitude the contributions of these otherwise uncredited readers

This book was set in ITC Stone and Avenir by New Best-set Typesetters Ltd. Printed and bound in the United States of America.

Library of Congress Cataloging-in-Publication Data

Names: Griffith, Saul, author.
Title: Electrify : an optimist's playbook for our clean energy future / Saul Griffith.
Description: Cambridge, Massachusetts : The MIT Press, [2021] | Includes
    bibliographical references and index.
Identifiers: LCCN 2020052159 | ISBN 9780262046237 (hardcover)
Subjects: LCSH: Electrification—Environmental aspects—United States. |
    Electrification—Economic aspects—United States. | Electric power production—
    Environmental aspects—United States. | Electric power production—Economic
    aspects—United States. | Energy development—United States. | Energy policy—
    United States. | Clean energy—United States. | Renewable energy sources—
    United States. | Climate change mitigation—United States.
Classification: LCC TD195.E4 G735 2021 | DDC 333.793/20973—dc23
LC record available at https://lccn.loc.gov/2020052159

10  9  8  7  6  5  4  3

Thank you, Arwen, for everything.

Especially for Huxley and Bronte, who give me hope and purpose.

This is an emergency as serious as war itself.
—Franklin D. Roosevelt

We're not alone. Good people will fight if we lead them.
—Poe Dameron, *Star Wars: The Rise of Skywalker*

Americans will always do the right thing—after exhausting all the alternatives.
—Winston Churchill

# CONTENTS

# PREFACE

In this book I approach the climate emergency from a new angle. I look for solutions, not barriers. Solving climate change should taste at least as good as carrots, at best ice cream, but it should not be painful. Instead, I'd like to offer a no-regrets pathway to success.

All too many people in climate advocacy or climate work are beginning with the question of "what is politically possible?" That could be a result of the frustration that drives many people, including our children, to march and protest for more rigorous climate action. But aiming only for what is politically possible is the art of limiting ambition before you begin.

This book doesn't start with the question of what is politically possible, but asks what is technically necessary to reach a climate solution that is also a great economic pathway for a country. After we realize what is technically necessary, America needs nothing short of a concerted mobilization of technology, industry, labor, regulatory reform, and, critically, finance. Every stakeholder needs to coordinate their efforts to create the lowest-cost, zero-carbon energy system for all citizens.

The book provides details about one probable pathway to total decarbonization. Because I am trying to paint a picture of the future that is complete and compelling, some readers might think I am "picking winners" among clean-energy solutions. This book attempts to be technology agnostic—but not at the expense of exploring the likely technological outcomes. Fusion would be great, and nearly free carbon capture would be useful, but I'm not here to champion specific ideas; instead, I support technologies that pass the "Is it ready and does it work?" test.

The pathway that works is best summarized as "electrify everything."

The book leans on real data, much of which was assembled in an unprecedented analysis of the US energy economy that I undertook under contract with the US Department of Energy. These details provide a story that is less about abstract concepts than about the recognizable technologies that define our world. This book provides a high-resolution picture of the consequences of electrifying everything. Will our lives change? The surprising answer is, not radically. Those things that will change are for the better: cleaner air and water, better health, cheaper energy, and a more robust grid. Our citizens can keep pretty much all of the complexity and variety promised by the American dream, with the same-sized homes and vehicles, while using less than half the energy we currently use. This is a success story that casts aside the 1970s-era narrative of trying to "efficiency" our way to zero emissions. Our country faces a challenge of transformation, not of deprivation.

How do we ensure the lowest cost of energy while electrifying everything? First, policymakers have to rewrite the federal, state, and local rules and regulations that were created for the fossil-fueled world and which prevent the US from having the cheapest electricity ever. Our country needs to massively scale up the industrial production of technological solutions, just as we did to win World War II. We cannot take our foot off the innovation gas—although I'll argue that we don't need any major breakthroughs, as thousands of little inventions and cost reductions are the key to achieving our end goal. Finally, we must have cheap financing for our transition to a zero-carbon energy system with low-interest "climate loans." Climate change will not be solved if only the richest 10% can afford it; we need mechanisms to bring everyone along for the ride. In our nation's history, there are precedents for doing this: the US pioneered public-private financing in the past. Innovative versions of this can help us get the job done today.

The consequence of getting the technology, financing, and regulations right is that every family in the US can save thousands of dollars each year.

We need to triple the amount of electricity delivered in the US. What is required is a moonshot engineering project to deliver a new energy grid with new rules—a grid that operates more like the internet. To do this, I argue that we must have "grid neutrality."

The industrial mobilization required to hit the climate targets that our children deserve will require an effort similar to World War II's "Arsenal of Democracy" in size, speed, and scope.

For a world desperate to rebound from a pandemic and economic crisis, there is no other project that would create this many jobs. I've worked with an economist to include an analysis that projects the creation of as many as 25 million good-paying jobs, spread across every zip code, suburb, and rural town in the country, should we choose to address climate aggressively.

This will not be easy, and people will tell you it is politically impossible. But, as I argue in this book, it is still possible. The earth is bigger than politics, and to meet our challenge, politics as usual must change.

Our future on this planet is in jeopardy. Billionaires may dream of escaping to Mars, but the rest of us . . . we have to stay and fight.

# 1

# A GLIMMER OF HOPE

- To eliminate all of our carbon emissions, the only serious option is to electrify (nearly) everything.
- To hit our climate goals, we need 100% adoption of electric solutions for households (electric vehicles, heat pumps, rooftop solar). This is your personal zero-carbon infrastructure.
- Massive generation and transmission infrastructure buildout are needed to decrease carbon emissions.
- New financing mechanisms—"climate loans"—are required so that everyone can afford to be part of the solution.
- Electrifying everything will require three to four times as much electricity. It needs to be generated, transmitted, and stored with "grid neutrality," where households, businesses, and utilities operate as equals.
- Fossil-fuel subsidies must be eliminated, along with rules and regulations that artificially inflate the costs of renewable energy and clean solutions.
- A wartime-like mobilization of industry is required to decarbonize on schedule for a 2°C/3.6°F increase in global temperature.
- A 1.5°C/2.7°F global temperature increase is now only possible if we promote negative-emission technologies and retire the heaviest emitters.

This book is an action plan to fight for the future. Given our delays in addressing climate change, we must now commit to completely transforming our energy supply and demand—"end-game decarbonization." The world has no time left.

A lot of people, including many politicians, activists, academics, and scientists, have given up. Sometimes I feel despair, too, given the widespread inertia and denial about climate change. But I refuse to give up. We have to fight not only the fossil-fuel interests but also the people who think we can't change our politics in time to save the future. As an engineer and an expert in energy systems, I can squint at the data and see a way forward to keep carbon emissions down to a point where the earth will remain livable and beautiful for future generations. If America does it right, every consumer will save money and the country will create millions of good new jobs and revitalize local economies.

In this book, I'm going to map out a viable path to averting a climate crisis. The path I lay out is not the only one available, but I can illustrate it in enough detail to reassure you that averting climate catastrophe won't require turning the world upside down. We have one last chance to address climate change, one glimmer of hope, and we must act now.

It's now time for end-game decarbonization, which means never producing or purchasing machines or technologies that rely on burning fossil fuels ever again. We don't have enough carbon budget left to afford one more gasoline car each before we shift to electric vehicles (EVs). There isn't time for everyone to install one more natural gas furnace in their basement, there is no place for a new natural gas "peaker" plant, and there is definitely no room for any new coal anything. Whatever fossil fuel machinery you own, whether it is as a grid operator, a small business, or a home, that fossil machinery needs to be your last.

My glimmer of hope comes from knowing that many of the barriers to a clean-energy future are systemic and bureaucratic, not technological. We have the technical means to address climate change, to have cleaner air and a verdant future without giving up our cars and the comforts of home. People have come to believe it will take a miracle to address climate change. It won't; we just need hard work! We have been told it will be too expensive, but doing it right it will actually save us money. Doubters say it will cost jobs, but embracing a green future will, in reality, create millions of them. Most people believe a clean-energy future will require everyone to make do with less, but it actually means that we can have better things.

There are obviously a lot of barriers to accomplishing this plan. I tell people what is technically necessary, and they tell me about political barriers. As naive or implausible as it may sound, we have to figure out how to remove all of those barriers—one at a time, and then, hopefully, many at once. Policymakers have to change what they believe is possible in the current economic and political climate. If what is politically possible is the extent of their ambition, everyone is doomed.

Fortunately, the younger people striking for climate change have not given up, and thank god for them and for others who are doing their part. This book is for those of you who have hope—and those who are willing to fight. I want this book to give you a blueprint of demands so that your entreaties to politicians may be detailed and your requests of business leaders specific. They failed to provide a road map to the future we want, so now you must give it to them, and urgently.

I have challenged myself to give you a very detailed answer to what is technically necessary, based on the best, most comprehensive data that can be compiled. If we know what is technically necessary, then we can get creative with the questions of how to make it politically possible and economically viable.

As my bioengineer friend Drew Endy sometimes quips, "For the first time in human history, we have the technology for nine billion people to prosper on this planet, but our politics and institutions haven't caught up."

Our leaders' failure to mitigate the COVID-19 pandemic certainly does not inspire hope that they are up to the task of tackling climate change. Rather than being prepared for this type of crisis—which scientists had predicted for many years—they spectacularly fumbled their response to the pandemic. COVID-19 has an important lesson to teach us, though. The curve in figure 1.1 that we had to bend to successfully deal with COVID-19 is the same shape as the curve we need to bend to successfully address climate change. Climate change, however, presents even more difficulties. Whereas with COVID we needed to act 20 days ahead of the virus, with climate change we need to act 20 years ahead.

## INFRASTRUCTURE LIMITS & BENDING THE CURVE

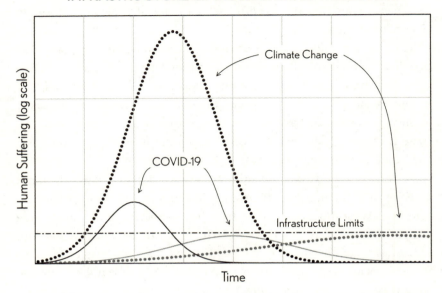

**1.1** Flatten the curve! Climate change is similar to COVID. It is necessary to act long before the worst effects of climate change are obvious. With COVID, action is required a few weeks in advance; with climate change, a few decades in advance. With COVID-19, the infrastructure limit is hospital beds; with climate change, it is our planetary life support systems.

Both are problems that require advanced preparation and science-based policy.

COVID-19 now has multiple vaccines, and many thanks to the scientists and engineers who made that happen. We also already have a vaccine for solving climate change. That vaccine is clean-energy infrastructure. We know what that looks like: massive electrification with wind turbines, solar cells, electric vehicles, heat pumps, and a much-expanded electrical grid with internet-like neutrality to glue it all together.

While this might come as a surprise, if policymakers commit to electrifying our infrastructure at the scale required, energy costs will decrease for all Americans. This is especially true if decisionmakers accompany the project with an appropriate set of financing mechanisms—loans, incentives, and subsidies—that will make the electrified future affordable for everyone. We have the clean-energy solutions we need to keep our levels of carbon emissions low enough to enjoy a clean, green, and prosperous future.

I still see a glimmer of hope. But to turn that hope into a reality for the future, we have to ask and answer some critical questions, which will be the focus of this book:

**What is the urgency?** Carbon-dioxide emissions from human activities have heated the earth to dangerous levels that will harm unimaginable numbers of people, ruin economies, spark wars and mass migrations, decimate species, and damage the environment. "Committed emissions," fossil fuels slated to be burned by machines that already exist, make the situation more urgent than is generally realized. For any chance of hitting our climate targets, we need an almost 100% adoption rate of decarbonized energy solutions, starting now. This means we need to immediately scale up ready-to-go solutions—and not hope for miracles or solutions we haven't developed, such as cost-effective technologies that suck $CO_2$ out of the air. I will discuss this in chapter 2.

**What can inspire us?** The plan I outline in this book may sound so audacious as to be nearly unachievable. Yet with climate change, that's where we find ourselves: having to achieve the impossible. By looking to historical examples of America taking on daunting problems and succeeding at them, we can begin to see the pathways that can turn the impossible into the inevitable, as I argue in chapter 3.

**How do we know what we know?** Over the past 40 years, government agencies and scientists have collected the information needed to address climate change. By understanding these highly detailed energy datasets, scientists know where and how to replace fossil-fuel energy with decarbonized sources, and how much energy we'll save in the process, as we'll see in chapter 4.

**How should we change our thinking about climate change?** Unlike previous energy crises, this isn't a problem that can be solved with increased efficiency and simple improvements to current systems; it requires transformation. Hidden in our historical energy-use patterns is great news: we can completely decarbonize without drastically changing our lifestyles or giving up the things we know and love, and we can do it with half of the energy we use today. The clean-energy future is just plain better, as we'll see in chapter 5.

**What do we have to do?** Electrify (nearly) everything. On the supply side, we need massive deployments of wind and solar (and likely some nuclear), which are already cheaper than natural gas and other fossil fuels for producing electricity. Hydrogen and biofuels won't be playing starring roles, except in certain applications (like biofuels for air travel). On the demand side, we need a huge roll-out of electric vehicles, heat pumps, and energy storage, as I explain in chapter 6.

**Where will our energy come from?** Our energy, for the most part, will come from the sun and other renewable-energy sources. People fear a future that they can't imagine, and to dispel these fears, in chapter 7 I will outline the basic physics of energy supply to paint a picture of how we will power the future cleanly.

**How will we make it work 24/7/365?** People don't like it when their lights go out, so how can we make sure this system provides the reliable energy we have come to know and love? In chapter 8, I will address this question.

**What is infrastructure?** Many people have an outdated concept of infrastructure and think that it applies only to roads, bridges, dams, and transmission lines, but this is insufficient to describe the new world we need to build. By recognizing our homes, cars, and heating systems as critical to a balanced energy infrastructure, we enable new ways to think about financing them. Consumers will also get relief from the daily grind of small decisions as they come to realize that their personal climate footprint is largely determined by a handful of infrequent decisions, as discussed in chapter 9.

**Can we afford to make the switch?** America can't afford not to switch to clean energy, for the planet as a whole and for our nation's economy. Unlike fossil fuels, renewable energy is cheap and getting cheaper. As I explain in chapter 10, when these technologies are scaled up, clean energy will basically be "too cheap to meter," as they used to say about nuclear energy.

**But will you save money?** When energy is cheap, everything is cheaper. I have built a model from the kitchen table out, to show how a clean-energy transition will affect every household's budget. Here, you will see concretely how getting clean energy right will save every consumer money on their energy bills, as I show in chapter 11.

**How are we going to pay for the transition?** Perhaps a better question is, "at what interest rate?" Because borrowing money is the way to finance climate infrastructure. All of the technologies for decarbonization have high up-front capital costs and low lifetime fuel and maintenance costs. America solved similar finance problems with the invention of auto financing in the 1920s, the modern 30-year government-guaranteed mortgage in the 1930s, and rural electrification during the New Deal. An analogous financial solution is required today, as I argue in chapter 12.

**How will we pay for the past?** Climate activists can fight the fossil fuel companies until the end of our lives, or Americans can come together, thank these companies for a century of service, and engage with them in the fight for our future. See chapter 13.

**How do we rewrite the rules?** We live with a legacy of regulations written for a fossil-fueled world. People broadly understand the problem with subsidies for fossil fuels, but more importantly, and less obviously, policymakers need to eliminate the regulations that artificially increase the price of doing the right thing. Our leaders need to write simple rules that encourage the best energy system, electric vehicles, and electrified buildings that America can make, as I urge in chapter 14.

**What about jobs and the economy?** The COVID crisis has caused the highest unemployment rate since the Great Depression. Like America's manufacturing efforts for World War II, our country can create new jobs with massive investment in infrastructure—this time, clean energy. If we make the switch to a decarbonized economy, we will gain millions of jobs, as I show in chapter 15.

**Can we handle this enormous challenge? Is there a precedent?** Industrial mobilization for World War II is the closest analogy we have to the scale, difficulty, and cost of solving the problem. In chapter 16, I provide a detailed breakdown of how that played out and I explore how to win this fight for our climate.

**Isn't climate just one of our many environmental problems?** Yes. Even if we solve climate change, you might say, the oceans will still be suffocated by plastic, the Amazon rainforest will still burn, and coral reefs will be decimated by agricultural run-off. In chapter 17, I will look at all of the tons of materials that flow through our lives. This will reveal opportunities not only for reducing our energy consumption and carbon output but also for sequestering significant amounts of carbon and reducing our larger footprint on the earth.

**What about carbon sequestration, carbon taxes, hydrogen, and other plans to fight climate change without electrifying everything?** There's too much carbon to sequester, it's too late for carbon taxes, and hydrogen is a false god. We need some of all of these things, but as we will find out in appendix A, they are not "get out of jail free" cards.

**How can you make a difference?** Everyone can contribute their personal efforts and skills to a war-scale mobilization effort. The only way we're going to win the battle against climate change is to keep fighting. Always demand more from political and business leaders. We lose the battle against climate change one compromise at a time. When politicians set targets for 2050, you need to demand targets for 2030. When industry says they will transition via natural gas, you need to reply that there is no more time for natural gas (and there's nothing natural about it). When people say that it doesn't matter what they do because China or Russia or India or Brazil won't do it, you need to respond that America will show other nations the way. The world can't afford any delays due to despair. That despair must be channeled into hope, and hope converted into action, as I argue in appendix B.

**Who am I?** I am a scientist, engineer, inventor, and father who wants to leave my kids a better world. I'd also like them to feel the sense of awe for our planet and its creatures that I have been lucky enough to enjoy. I am in this fight and I'm giving it all I've got. The data convinces me that it is still rational to have hope—but not for much longer. We can win big against this climate emergency, but this is our last chance. If we win——when we win, because there is no other option—we'll all be much better off than before.

## IT'S A CLIMATE EMERGENCY. PLUG IN. ELECTRIFY!

This book is principally concerned with the emergency of the nearly 75% of greenhouse-gas emissions related to the US energy system, which

# Millions of Tons of CO$_2$ Emissions by Sector and Type

☐ Waste, landfill, 134 ☐ Industrial emissions, 376
☐ Agriculture, 618 ☐ Energy sector emissions, 5,547

Landfill | Wastewater

Soil, fertilizer | Livestock | Manure | Rice

Refrigerants (A/C, refrigeration) | Steel | Cement | Petro-chemicals | Ammonia | Lime

Natural gas supply chain | Fossil fuels as materials | Oil supply chain | Coal supply chain

Combustion of fossil fuels

**1.2** This book is principally about the biggest component of our CO$_2$ emissions: combustion of fossil fuels in the energy sector. From EPA estimates of US greenhouse-gas emissions. "Negative emissions" from land use not shown.

accounts for the overwhelming majority of our emissions (the US is representative of the global problem, so throughout this book, while we focus on the US, our analysis is usually a reasonable proxy for the entire globe).[1] Other emissions come from the agricultural sector (around 12%), land use and forestry (7%), and industrial non–energy use emissions (7%). Mobilizing to address climate change as suggested in this book would also address much of the industrial non-energy emissions, and a little of the other two, as well. Decarbonizing America's energy supply is about 85% of what we need to do. I have to believe that if we commit to solving 85% of the problem, the smart and passionate people working on the other 15% will do their part, too. For this reason, emissions unrelated to energy will receive only periodic mention throughout the rest of the book.

# 2

# WE HAVE LESS TIME THAN YOU THINK

---

> ☞ Climate change is more of an emergency than most people realize.
> ☞ Most commonly reported emissions trajectories assume we'll achieve rapid "negative emissions" later this century by pulling $CO_2$ out of the air. This is not yet viable. We cannot rely on miracles.
> ☞ Committed emissions—fossil fuels slated to be burned by machines that already exist—make the crisis even more urgent.

With climate change, the science is clear. Scientists have written a large body of work on global warming and can predict the future climate from estimates of our current carbon emissions. We know, with certainty, that we are hurtling toward multiple environmental and human catastrophes. (See the primer on climate science in appendix C.)

We can no longer debate the science. For some people, science-based arguments will never be enough. The scientific theory of evolution has existed for more than 150 years, with irrefutable evidence, yet only about 35% of Americans believe that we evolved by natural processes.[1] In late 2019, I visited my friend Louise Leakey in the Rift Valley of Kenya, where early humans evolved. Her family has been studying the origins of human evolution for generations. As Louise pointed out the features of these million-year-old skulls to my six-year-old daughter, it demonstrated the obvious—there really isn't much room for doubt.

For those who likewise doubt the science of global warming, there are other reasons to support efforts for a zero-carbon future: it will likely save us all money, improve the overall economy, clean our air, and improve our health. Still, whatever evidence we deploy, it's likely we'll have to solve climate change without broad consensus, because culture moves more slowly than science.

Yet more and more influential people across the globe and the political spectrum have realized we are in an emergency: the Pope;[2] the Dalai Lama; most Democratic leaders and Republicans including Senators Mitt Romney, Mike Braun, and Lindsey Graham;[3] young activists such as the Youth Climate Strikers; older activists like Extinction Rebellion; and young Democrats and Republicans alike.[4] Polls show that the majority of Americans believe that the government isn't doing enough to protect the climate and environment.[5] Establishment figures such as Christiana Figueres, the former Climate Chief of the UN, are even calling for civil disobedience.[6] Jane Fonda has already been arrested many times for the cause.[7]

Whether you think climate disasters qualify as an emergency may depend on where you live, how hot it's getting, and how high the seas are rising around you. My opinion, and the opinion of practically all scientists, is that it is definitely an emergency.

- If you are an Australian (like me), the fires, floods, loss of life and wildlife, and droughts caused by one degree of warming are already devastating: in the January 2020 bushfires, an estimated 25 million acres burned, killing one billion animals and two dozen humans. Coral reefs are already dying. The effects of 2°C/3.6°F would be terrifying.
- If you are a Californian (also like me), you'll see more megafires, causing deaths, damage to property, displacement, and air pollution.
- If you live on a low-lying island or a floodplain with a few hundred million people, like Bangladesh, 1.5°C/2.7°F is difficult and 2°C/3.6°F would be devastating, bringing flash floods, rising waters, contaminated water, disease, and the widespread loss of lives and homes.
- If you live in a low-lying city like New York, your city may build levies and breakwaters to tolerate the sea-level rise implied by 2°C/3.6°F, but big storm surges will still cause flooding. And those levies represent money not spent on other things.

- If you live in Miami or the Florida Keys, it is more likely that 2°C/3.6°F completely changes your beaches, and further sinks property (and property values).
- If you live somewhere like inland Canada or Russia, 3°C/5.4°F might not seem so bad and may even improve your agriculture, but that doesn't acknowledge the pressures you will feel in a world of hundreds of millions of climate refugees and the conflicts created by stressors on the global food system.
- If you are one of the roughly one-third of species that is threatened by extinction by climate change (including the bees and other pollinators on which human food supplies depend), you probably agree that no warming at all would be best.
- If you are a farmer, you are already dealing with changing weather patterns, seasons, and the viability of your crops.
- If you are an insurer, you may now be refusing insurance to clients for rebuilding after climate-related events, knowing that they will happen again.
- If you work in the medical field, you understand that climate change is a public-health issue akin to a pandemic, and that it will cause future pandemics. These effects already kill thousands of people per year and cost trillions of dollars in health care,[8] and the effects are getting worse every year.
- If you are a child born today and will live until 2100, when the projected sea-level rise is 2–10 feet, enough to displace hundreds of millions of people, you may not be able to say the words yet, but your future self knows you were born into an emergency.
- If you are in the military, you have already identified climate change as the biggest threat to national security because it will lead to more refugees, diminished supply chains, and the transition from small regions of instability to global instability.

It would be easy to write another doomsday book on climate change. Instead, I am going to show you a clear path to a better world in enough detail to bridge the imagination gap. This is where my hope is, based on science and what is technically possible.

But first, let's look at why the timeline for action is more urgent than you might think.

## WE MUST ACT NOW

It has to be now—not 10 years from now, or even a month from now. We have arrived at the last moment when we can shift global energy infrastructure without passing a 1.5°C/2.7°F–2°C/3.6°F temperature rise. We still have the opportunity to address climate change in a way that will make the future better.

The 2016 Paris Agreement aimed to avert climate crisis by keeping global temperature rise this century to 2°C/3.6°F above pre-industrial levels while pursuing efforts to limit the temperature increase even further, to 1.5°C/2.7°F.[9] The 1.5°C/2.7°F and 2°C/3.6°F targets were as political as they were technical, and in some respects were chosen simply because they are round numbers. The choice to express climate change in Celsius has been a challenge—a narrative problem that persists in the US, where a few degrees Fahrenheit doesn't sound too bad, which is why in this book we list the Fahrenheit and Celsius targets.

Even with the emissions targets championed in these agreements, we have a significant chance of failing to attain the climate stabilization we would like. In 2018, the Intergovernmental Panel on Climate Change (IPCC), a group of United Nations scientists who summarized the worldwide findings on climate change, concluded that meeting the Paris target of 1.5°C/2.7° F would be possible, but it would require "rapid, far-reaching and unprecedented changes in all aspects of society."[10]

The report predicted that "we have 12 years" to act if we want to reach this target. The report was issued in 2018, but we didn't really do anything to improve the situation in 2019 and 2020, so now we have 9 years to halve human emissions by 2030. The IPCC warned that even keeping warming to 1.5°C/2.7° F—already an ambitious goal—would result in large-scale drought, famine, species die-off, the loss of entire ecosystems, and the loss of habitable land, and would throw more than 100 million people into poverty, particularly in the Middle East and Africa.[11]

That's especially true because the IPCC report relied on humanity developing "negative emissions" technologies, such as carbon sequestration, to reach that goal. But at the moment, those technologies don't yet exist on a workable scale, and there are strong indications that they will never be cost effective.[12] We can't rely on fantasy technologies to reach our climate goal (or to argue that we can continue to burn fossil fuels because someday we may be able to suck the $CO_2$ out of the air). We must aim to hit 2°C/3.6°F with technology that works *today*. Current technologies will do the job, if we employ them right away.

If we exceed our emissions targets, we will face irreversible tipping points that will make it impossible to stabilize the climate. As Timothy Lenton and his colleagues highlight in their recent paper, the more we learn about these tipping points, the more we understand that they will happen sooner, and with more disruption, than we had previously thought.[13] Given what we know about climate feedback and sensitivities—such as more rapidly melting glaciers, the effects of deforestation of the Amazon, methane emissions from Arctic tundra, and carbon releases from fires—we are already precariously close to such a tipping point. Some scientists argue that we've already lost Greenland's ice sheet.[14] Every year we wait—whether hoping for a political revolution or a technological miracle—has dire consequences to the health of our planet. This climate-response emergency is expressed best in the analysis and charts of Zeke Hausfather[15] and Robbie Andrew,[16] which we redraw in figure 2.1.

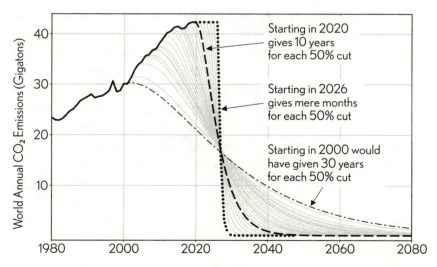

**2.1**  Mitigation curves required to hit a 1.5°C/2.7°F world, redrawn from Robbie Andrew's data. As this figure demonstrates, there is no time left to begin reducing emissions. If we don't, any chance of hitting the necessary climate targets will slip beyond our reach.

Here's how to look at this chart. If we had started this grand project in the year 2000, we could have hit our 1.5°C/2.7°F target by reducing emissions at the rate of 4% per year. If we start now, in 2021, we have to reduce them at a breakneck pace—something like 10% per year. If we wait four more years, we will use up half of the remaining carbon budget. In eight years, it's gone completely. We simply must start yesterday, or as my friend Jonathan Koomey says, we must "halve emissions every decade." I think we have to do even better.[17]

## COMMITTED EMISSIONS

The notion that we have 10 years also fails to recognize "committed emissions," those that are locked in because we have already invested in infrastructure that will emit carbon dioxide throughout its useful life. An example is the car sitting in your driveway that burns gasoline but is too new to trade in for an electric vehicle.

Fossil-fueled power plants built today will emit $CO_2$ for 50 years or more unless we shut them down. A gasoline-powered car or gas furnace purchased yesterday will probably discharge $CO_2$ for 20 more years. These committed emissions already take us past 1.5°C/2.7°F of warming and closer to the edge of 2°C/3.6°F.[18] That should sober us up, because it means that even if we made perfect climate decisions on every purchase from now on, we would still shoot past our 1.5°C/2.7°F target.

Let's reflect on what we have just learned for a moment: we have started this fight so late in the game that now every time we retire a fossil fuel-burning machine, it must be replaced with a decarbonized machine. This applies to everything and everyone that uses energy, whether an individual, a power company, or a corporation; all will require a decarbonized solution. In theory, this calculus would change a little if the country were to retire the heaviest-emitting coal plants before their end of lives. But that does not substantively change the fact that America needs to eliminate all fossil-fuel burning machines.

## 100% ADOPTION RATE

This scenario of replacing everything that uses energy with a zero-carbon solution when it's retired is called a 100% adoption rate. Today, when

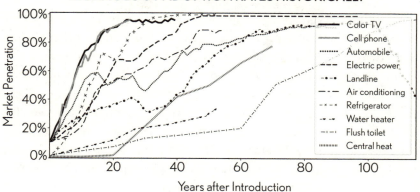

### TECHNOLOGY ADOPTION RATES HISTORICALLY

(Legend: Color TV, Cell phone, Automobile, Electric power, Landline, Air conditioning, Refrigerator, Water heater, Flush toilet, Central heat)

X-axis: Years after Introduction — Y-axis: Market Penetration

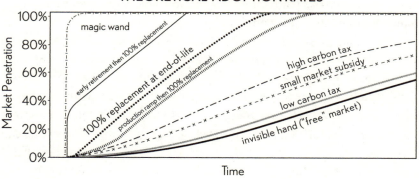

### THEORETICAL ADOPTION RATES

magic wand; early retirement then 100% replacement; 100% replacement at end-of-life; production ramp then 100% replacement; high carbon tax; small market subsidy; low carbon tax; invisible hand ("free" market)

X-axis: Time — Y-axis: Market Penetration

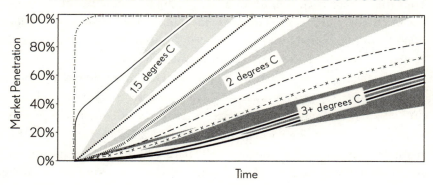

### THEORETICAL ADOPTION RATES VS. CLIMATE OUTCOMES

1.5 degrees C; 2 degrees C; 3+ degrees C

X-axis: Time — Y-axis: Market Penetration

**2.2** a) Historical rates of technology adoption. Notice that even quickly adopted innovations (like cell phones) take 20 years to saturate the market. b) Qualitative adoption rate scenarios for various market drivers. "Free" market approaches are too slow. We need 100% replacement to meet our climate goals. c) Adoption rate scenarios overlaid with climate outcomes. We need 100% replacement at end of life to meet our climate goals.

a car reaches retirement age, there is only a small chance that it will be replaced by an EV. If 1 in 10 people buys an EV, then we say the adoption rate is 10%. Because machines like your car have long lifetimes, that means that traditional gas-powered cars will remain on the road for a long time. To reduce emissions, though, our world can no longer afford those slow adoption rates. We need everyone buying electric vehicles. Likewise, we need every corporation purchasing a power plant to choose solar instead of natural gas and wind instead of coal. Fortunately, we are further along with this project than you might expect. In 2018, 66% of new power plants globally were renewables or carbon free![19] But while this is good, it is not quite enough—across the board we now need adoption rates of 100%. This complete adoption rate is required by the endgame decarbonization we ultimately need.

While that sounds dramatic, it doesn't mean you have to run out to buy a new EV today. It means that the next time you need to retire a car or any other machine, it should be replaced with one that doesn't emit $CO_2$. When your car finally dies, you should replace it with an electric one. *Consumer Reports* says the average life expectancy of a new car is eight years and 150,000 miles of travel, though well-maintained cars can last much longer—I have a 1963 Land Rover with 400,000 miles on it and a new engine, but the next engine will be electric, even for that old jalopy. The same logic applies to your water heater, your furnace, and your stove. Your roof, too, needs a solar upgrade. Similarly, the natural-gas electricity generation plant that was built in your town in the mid-2000s won't be retired tomorrow, but it needs to be at the end of its life, which is probably 2040 or 2045. Start lobbying today.

Water heaters last 10 years. Refrigerators, 12; clothes dryers, 13; rooftops, 15; furnaces, 18; cars and trucks, 20; thermostats, 35; power plants, 50.[20] No matter how effective climate activists are at convincing people to buy green technology, we are unlikely to decarbonize faster than the natural lifetime of existing machines. That's why we'll need incentives such as buy-back programs and subsidies to swap out fossil fuel–burning machines for electric ones as soon as possible.

We can buy ourselves a little extra time if we shut down the most polluting infrastructure before it ends its natural life. This is why people advocate for early retirement of fossil-fuel power plants, particularly those that burn coal. But consumers, utilities, and other organizations will require extreme motivation to retire their fossil fuel–dependent infrastructure early because of their sunk costs. You aren't going to give up your gasoline-burning car unless there are financial incentives to make it easy for you to replace it with a new EV.

A 100% adoption rate is only achieved by mandate—and robust financial incentives to back it up. It typically takes decades for a new technology to become dominant by market forces alone as it slowly increases its market share each year. Electric cars still only represented 2% of sales of US vehicles in 2018, though they represented 5% of all vehicle sales in California in 2019; but this is 15 years after Tesla was founded and 20 years after GM shut down the production of its first electric car, the EV1. We need EVs and other emissions-free vehicles to be 100% of vehicle sales as soon as is physically, and industrially, possible. We don't yet build even one million EVs per year in the US, and the new vehicle market is 17 million cars, trucks, SUVs, and minivans per year.

The challenge of 100% adoption presents a giant conflict that we need to address head-on: the "free market" as we know it is not up to the task of keeping the world below 2°C/3.6°F and has absolutely no chance of allowing us to hit 1.5°C/2.7°F. It may sound like this is a screed for government intervention, but it isn't! I am merely stating what is technically necessary. If your toilet was broken and you called me and asked me what to do, I wouldn't tell you "the free market will fix that," I'd tell you to call a plumber. That is where the world is when it comes to climate change: no amount of hope in free-market solutions can change the fact that it is now too late to rely on the free market to act fast enough. We need to call the plumbers (and electricians, and engineers, and manufacturers) to fix our infrastructure now.

This is not to say that businesses and the market don't have roles; they are critical. But in emergencies, ideologies must be put aside. When Mother Nature arm-wrestles with the invisible hand, she will always win. As my friend, the economist Skip Laitner, says, the free market needs an invisible foot to give it a swift kick in the ass now and then. It is urgent for every player to act and do their part. Individuals, governments, businesses, and the market—we need every tool in the box, and we need them working together.

As we will discuss in the following chapters, the emergency response to this climate emergency is fairly simple, in concept:

- We must electrify the vast majority of our energy supply and uses. That electricity must come from renewables and nuclear power.
- We must transform our heavy infrastructure as well as the personal infrastructure we create with our household purchasing decisions.
- Your next car needs to be electric. Your next furnace needs to be a heat pump. You need solar on your roof. This is your personal decarbonization infrastructure.
- We must demand that politicians drive this transformation faster than free-market forces alone are capable of doing.
- Industry must be incentivized to ramp up production of green technologies at a rate similar to wartime mobilization.
- Bankers and policymakers need to create new financing mechanisms so that everyone can afford to be part of the solution.

Decarbonizing our country and switching to clean energy will create jobs in every zip code—in manufacturing, construction, installation, infrastructure, agriculture, and forestry. This is a chance to revitalize our cities, rejuvenate our suburbs, and reignite our rural towns. We can rebuild a prosperous and inclusive middle class, as we enjoyed after World War II, with tens of millions of good new jobs that are vital and proud. If America does it right, everyone's energy costs will go down. Everyone has a role to play in the war effort.

We now face a climate emergency as challenging as all of our other twentieth-century emergencies combined. It requires mass mobilization with extraordinary speed and resources. Without a doubt, you are worried, scared, or worse. That's reasonable, but we can't do nothing, and as I argue in the next chapter, this is also a vast opportunity to make the world, and our economy, better for everyone.

# 3

# EMERGENCIES ARE OPPORTUNITIES FOR LASTING CHANGE

---

> ☞ Prior emergencies America has confronted provide examples of what we need to do to boldly avert climate change.
>
> ☞ The US, somewhat uniquely, has a track record of stepping up to the plate in an emergency.
>
> ☞ Bold action in the face of a crisis can make lasting improvements to our quality of life.

Despite the US's botched response to the COVID-19 pandemic, this country has successfully fought many other emergencies throughout its history. Americans have made a difference through individual and collective actions. Whether the threat was to the wilderness, prosperity, democracy, civil rights, technological superiority, national security, public health, or the hole in the ozone layer, in each case, the US faced a strong enemy—and won. For inspiration and guidance, it's worth taking a moment to reflect on how we overcame these obstacles. We can also look at past challenges to understand the tools we can use from history to help fight our climate crisis.

## SAVING THE WILDERNESS

In 1903, the naturalist John Muir realized that many of America's wild lands[1]—"temples of nature," he called them—were being stripped for logging, mining, and development. If the destruction continued, those wild places would be gone. They urgently needed to be protected before they were forever destroyed. Muir convinced President Theodore Roosevelt to come camping with him in Yosemite, roughing it for three days while he impressed upon the president the need to protect public lands to preserve America's natural resources for future generations. (Imagine a president who would go camping as an example for the nation, instead of playing golf!) It worked: during his presidency, Teddy Roosevelt signed into existence 5 national parks, 18 national monuments, 55 national bird sanctuaries and wildlife refuges, and 150 national forests.[2] This caused the displacement of Native Americans; but we can also celebrate Roosevelt's vision and tenacity in preserving wilderness for future generations.

**We have it in us to preserve our natural world for future generations to enjoy.**

## THE NEW DEAL

Between 1933 and 1939, President Franklin D. Roosevelt, working with Congress and advisors, enacted a series of jobs programs, public works projects, and financial reforms to help Americans recover from the Great Depression. One was the modern long-term, government-backed home mortgage, which allowed many people to buy homes and anchored a stable, enduring middle class. These programs helped millions of Americans, but too many people were unjustly left out. African-Americans, for example, were excluded from the housing market and federal mortgages.

Today, America has the once-in-a-lifetime chance to solve the current economic crisis. Unlike the case of the New Deal, we can do so inclusively and equitably, while at the same time confronting the impending climate disaster and decarbonizing the country. Mortgages and low-interest loans are important in the context of the climate emergency, because while clean-energy sources produce almost free electricity when they're up and running, they require up-front cash. You have to have the spare capital to put solar panels on your roof in order to enjoy the long-term savings they offer. Fixing the climate will require "climate loans" that will make it easier to buy electric cars and electric home-heating units rather than continuing to rely on fossil fuel–powered machines.

Another New Deal program that can serve as a model for financing electrification is the Rural Electrification Act of 1936, which provided federal loans to install electrical systems to rural areas in the US. The Electric Home and Farm Authority (EHFA) helped rural Americans finance purchases of electric appliances such as refrigerators, ranges, and hot water heaters. EHFA ultimately financed some 4.2 million appliances, at a time when there were around 30 million households in the US.[3]

**Innovative financing plans can pull us out of a crisis and build a strong basis for a more prosperous citizenry.**

## THE MOBILIZATION FOR WORLD WAR II

After Hitler's troops marched into France, and after Britain had to retreat from Dunkirk, the situation in Europe—and the future of democracy—looked dire. Winston Churchill, flailing against Hitler, entreated Roosevelt to join the war. Roosevelt responded by creating an industrial infrastructure capable of out-manufacturing Germany in a new type of war that would be won not just with soldiers, but with airplanes, tanks, jeeps, guns, bullets, boats, and bombs. The US was initially in no shape to take that on. Coming out of the Depression, the country was in an isolationist mood, and the military was under-equipped and disorganized. Roosevelt partnered with industrialists to build the armaments we needed to get the job done in record time.

**We are capable of ramping up industrial production at an astonishing rate—fast enough to make the necessary technological changes to meet the crisis.**

## THE SPACE RACE

On October 4, 1957, the Soviet Union surprised President Dwight D. Eisenhower and the US by successfully launching Sputnik I, the world's first artificial satellite. The beach ball–sized Sputnik set off the US-USSR space race and launched new political, military, technological, and scientific developments.

Immediately after Sputnik, the US created a series of nimble science agencies to avoid future surprises and to chart a path forward, including the National Aeronautics and Space Administration (NASA) and the Defense Advanced Research Projects Agency (DARPA). (DARPA started out as ARPA, and the D—for Defense—was added in 1972.) These agencies have gone on to make astounding technical advances in artificial intelligence, stealth technologies, microelectronics, surveillance, and communications—including the prototypical communications network ARPANET, which evolved into the worldwide internet as we know it today.

President John F. Kennedy leveraged Eisenhower's agencies to launch a technical project so ambitious that it now defines scientific and engineering ambition: the moon shot. On March 25, 1961, he declared a dramatic goal: to land an American on the moon within the decade. On July 20, 1969, Apollo 11 landed on the moon—Neil Armstrong's small step and "giant leap for mankind." The space effort gave humanity a vision beyond our own tiny planet and helped us see ourselves as but one species in the larger context of the solar system and the universe.

In today's dollars, the Apollo program cost $150 billion over its 10-year lifetime. Currently, the US government only spends about $3 billion annually on energy and climate technologies—approximately one-fifth of the rate of spending of the moonshot. The Department of Energy has a budget of around $30 billion, the great majority of which is spent on nuclear deterrence, arms stockpiling, and security. The DOE invests heavily in fundamental science, but only a small fraction, around $3 billion, is invested in energy technologies that are likely to make an impact in the near term.[4]

Since we're talking about saving the planet, a 10- or 50-fold increase in energy-technology spending seems reasonable.

**We can invest massively in science and technology to solve audacious problems.**

## THE CIVIL RIGHTS MOVEMENT

The civil rights movement fought the deeply rooted human emergency of institutionalized racism in the US. A succession of courageous activists, from Rosa Parks and the Freedom Riders to those who participated in the 1963 March on Washington, in which Dr. Martin Luther King, Jr. proclaimed "I have a dream" for racial equality, helped change discriminatory laws. King was assassinated, and the civil rights movement had to fight against dogged opposition across the country, but the movement was responsible for pushing Lyndon B. Johnson to pass the Civil Rights Act in 1964, the Voting Rights Act in 1965, and the Fair Housing Act in 1968. Since then, we've seen rollbacks to voting rights, but America also elected its first Black president, Barack Obama, and has seen other gains in diversity and inclusion. The Black Lives Matter movement, created in response to racist police brutality, has awoken many Americans to the persistence of discriminatory policing and violence against people of color. Civil rights activists have been—and continue to be—a model for many activists, including climate activists and the youth who are rising up and demanding their right to a livable future. Today's climate activists understand how the devastating global effects of climate change disproportionately affect people of color.

**People, together, can change the course of history with their collective activism. It requires bravery and direct action.**

## THE 1973 ENERGY CRISIS

Late in 1973, President Richard Nixon addressed the nation about "the energy emergency," issuing a warning about our reliance on foreign oil. The energy crisis demanded an ambitious response from US policymakers. President Nixon created science-based agencies to study and solve environmental problems: The Energy Information Administration (EIA), the Department of Energy (DOE), and the Environmental Protection Agency (EPA). Much of our understanding of our energy and climate crises is in the wheelhouse of these agencies that were shepherded into existence through three consecutive presidents: Nixon, Ford, and Carter.

Back then, the problem was that we were importing 10% of our energy from foreign sources, so we could reasonably count on figuring out how to use fossil fuels 10% more efficiently to solve the problem. That's how we got CAFE efficiency standards and Energy Star appliances. But this also left Americans with a now-outdated sense that we can solve energy problems with efficiency alone. While the 1970s energy crisis was about the 10% of our energy system that used imported oil, the current crisis is about transforming nearly 100% of our energy system to clean electricity.

Today, we need to stop using fossil fuels altogether; we can't "efficiency" our way to carbon zero.

**We understand our energy needs and strategy now because America pioneered the comprehensive collection of energy data in the 1970s. We need to invest further in our existing federal technology innovation system and data collection to develop the technologies we need to get to carbon zero, at scale and on time.**

## SMOKING, A PUBLIC HEALTH CRISIS

In 1964, US Surgeon General Luther Terry dropped a bombshell on the American public: smoking causes lung and other cancers, and the tobacco industry misled consumers by hiding the dangers of cigarettes. At the time, 42% of adult Americans smoked regularly. The Surgeon General mounted a public campaign against smoking that included health warnings, advertising bans, and a public awareness campaign to alert the citizenry to smoking's dangers.[5] Since then, the percentage of smokers has dropped by more than half, to 18%. The *Journal of the American Medical Association* estimated that over that period, our crisis response to the smoking epidemic prevented eight million deaths.[6]

Climate change also poses a grave danger to human health. The World Health Organization has estimated that meeting the goals of the Paris Agreement could save seven million lives worldwide each year by 2050 by reducing air pollution, which causes asthma and other respiratory illnesses.[7] The EPA estimates that the higher concentrations of ozone in the air due to climate change may result in tens of thousands of additional ozone-related illnesses and premature deaths per year by 2030 in the United States.[8] Global warming will also result in increased heat strokes and other heat-related deaths.

**A concerted public effort can avert a public-health crisis and rein in companies that promote ill health, whether Big Tobacco or Big Fossil.**

## OZONE DEPLETION AND REFRIGERANTS

After scientists discovered the large hole in the ozone layer, which protects us from harmful UV radiation, nations came together to agree to the Montreal Protocol in 1987.[9] They signed an international treaty to phase out the chlorofluorocarbons (CFCs) that were in most refrigerants at the time. We have amended the Montreal Protocol many times, including the recent Kigali Amendments. That probably wasn't attributable solely to altruism—Dow was making less money off CFCs in the 1980s, so it started to support the Montreal Protocol to phase out CFCs in favor of hydrofluorocarbons (HFCs), which it has a patent on.[10] Now, in the 2020s, the same story is repeating itself, with chemical companies, such as DuPont, Chemours, and Honeywell, funding the Kigali Amendments, which phase out HFCs, because they have new patents on hydrofluoroolefins (HFOs).[11] They're also trying to resist deployment of natural refrigerants which are competitors to HFOs.[12] Despite this industry-insider mischief, lowering the emissions of refrigerants is a great example of international cooperation in the face of a global emergency. I mention heat pumps frequently in this book; just like refrigerators and air conditioners, they use refrigerants and could be disastrous for the atmosphere if it weren't for the fact that science has already figured this out. The future of refrigerants involves "natural" refrigerants like supercritical $CO_2$ that have comparatively miniscule greenhouse-gas impacts.

**Nations came together to stabilize a complex geological system through collective action. Science identified the problem, engineers created solutions, and politicians created the right regulatory environment.**

## TODAY'S CLIMATE EMERGENCY

- Similar to the creation of the national parks, America has an opportunity to save beautiful wild places—and the whole planet—for our children.
- Like the New Deal, this crisis will require innovations in financing and public-works projects, and it will create employment.

- Like the World War II mobilization, America must turn to industry to transform infrastructure and accelerate the wartime production we need to solve an urgent problem. If not done voluntarily, this may require federal mandates through emergency powers.
- Like the Space Race, the country must commit to ambitious timelines and massive investments in science.
- Like civil rights, the legal response must be supplemented by direct action and social movements that create the political pressure for change.
- Like the 1970s energy crisis, data must guide our actions.
- Like the public health crisis that is smoking, we must use a combination of incentives—regulation, pricing, public awareness, and availability—to decarbonize.
- Like the Montreal Protocol, America should lean in to international policymaking that will address this crisis.

But the climate crisis we face today is in many ways different from these previous crises. This time the enemy—fossil fuels—is integral to our existing economy. This time, because of the lag-time in climate response, we need to act long before the worst impacts are felt. It is for these reasons that climate change has been described as a "super wicked hard problem"—problems that have been defined as a special category of almost impossible tasks.

Our reward for this work—besides saving the planet—will be abundant cheap energy, quality jobs, improved public health, and a new era of prosperity. We must be bold again.

# 4

# HOW DO WE KNOW WHAT WE KNOW?

> - The great data we have today was made possible by the civil infrastructure we built in the 1970s to collect that data.
> - The 1970s oil crisis could have been solved with energy system efficiencies.
> - Today's climate crisis is different from the oil crisis and must be solved by transforming the energy system.
> - We must decarbonize demand with the same urgency that we decarbonize supply.

The climate crisis is clearly an emergency. What expertise can we call upon to solve this emergency? We need to know where our energy is currently being supplied and used so that we can substitute cleaner sources of energy, and cleaner end uses of energy, to get to a carbon-free future.

The knowledge we have about our energy sources and usage comes from our last energy crisis, in the 1970s. Since then, we have accumulated mountains of data about our energy supplies and demands. But because that was a different kind of crisis, we have legacy ideas about energy that we need to shift before we can begin the work of getting to carbon zero.

The 1970s crisis was an oil imports crisis. It was a supply crisis, since about 10% of America's energy use—the oil from the Middle East—had been cut off. Since supply must equal demand, experts looked at the demand side—how we used energy—and found that we could easily be 10% more efficient in our use, in particular with our cars and appliances, and thereby eliminate the need to import fuels. Efficiency can solve a problem that relates to 10% of supply. This is what gave us Corporate Average Fuel Economy (CAFE) standards and Energy Star appliances. But, as we've seen, in our current moment we need to get to carbon zero, and you can't efficiency your way to zero. Your efficient gasoline powered car doesn't get us to zero emissions unless you **never** drive it. America has a new kind of energy crisis. So while we have legacy tools to understand our energy supply and demand, we need to update those tools, as well as our thinking, to meet the challenge of the current climate crisis.

## THE ORIGIN STORY OF ENERGY DATA

Late in 1973, Americans faced long lines and rising prices at every gas station. Energy was on everyone's minds, no matter their political leanings. Public interest in energy issues in the 1970s was so high that those beloved coal-burning cave people, Wilma and Fred Flintstone, starred in a TV special, *Energy—a National Issue* (figure 4.1), narrated by Charlton Heston (who went on to become a five-term president of the National Rifle Association). The equivalent today would be an entire *Simpsons* or *South Park* episode featuring Clint Eastwood that would educate the public about how to address climate change.

Congressman Melvin Price, then chairman of the Joint Committee on Atomic Energy (an intellectual precursor to the Department of Energy), tasked his staff with creating a comprehensive energy audit. He ordered them to take all the known energy-use data for the United States and prepare it as a display, which, "In less than an hour, could give an extremely busy person an understanding of the size and complexity of our national energy dilemma."

Working for the committee, Jack Bridges, director of National Energy Programs, Center for Strategic and International Studies at Georgetown University, devised a highly detailed Sankey diagram to map US energy use, and followed this up with his groundbreaking book, *Understanding the National Energy Dilemma*.[1] (See appendix D for details on how to read a Sankey diagram.) Bridges's diagram explained how we produced and

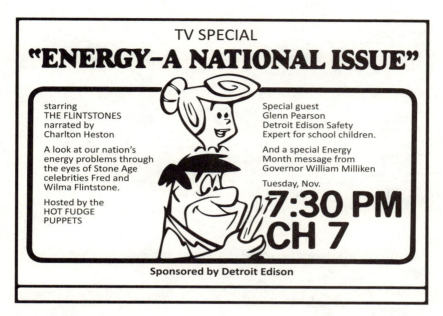

TV SPECIAL

# "ENERGY–A NATIONAL ISSUE"

starring
THE FLINTSTONES
narrated by
Charlton Heston

A look at our nation's
energy problems through
the eyes of Stone Age
celebrities Fred and
Wilma Flintstone.

Hosted by the
HOT FUDGE
PUPPETS

Special guest
Glenn Pearson
Detroit Edison Safety
Expert for school children.

And a special Energy
Month message from
Governor William Milliken

Tuesday, Nov.

7:30 PM
CH 7

**Sponsored by Detroit Edison**

**4.1**  *Energy—a National Issue*, as seen in your TV guide. Source: WXYZ-TV, *TV Guide Magazine* (Detroit Edition), November 19–25, 1977.

used our energy. The introduction was blunt: "The United States, with 6% of the world's population, was consuming over 35% of the planet's total energy and mineral production."

Bridges's flow diagram detailed America's use of oil and natural gas and showed how much electricity the country produced and how efficiently, with breakdowns of industrial use vs. that of commercial, residential, and transportation sectors. This work influenced the way we would measure and summarize energy data for decades to come. The left-hand side of the diagram is the supply—where the US gets its energy. The right-hand side is demand—what we use energy for.

The 1970s oil crisis actually resolved itself before the country had meaningfully improved the efficiency of its car fleet—and before we meaningfully changed our consumption behaviors or where we obtained our energy. Indeed, in figure 4.2 I compare the 2019 Sankey diagram to the first one published by Lawrence Livermore National Lab (LLNL) in 1973. To this day, Lawrence Livermore publishes Sankey flow diagrams each year on the data collected by the Energy Information Administration[2]—I

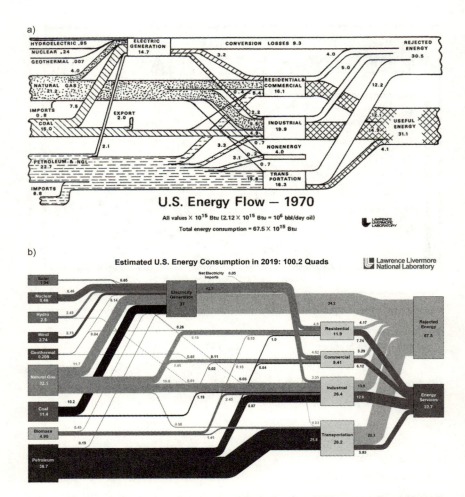

**4.2** a) The first LLNL Sankey diagram I can find, showing 1970 energy flows. b) LLNL Sankey from 2019, which looks largely the same, except a greater proportion of energy is "wasted" (although this is largely a difference in methodology). Source: Lawrence Livermore National Laboratory, "Energy Flow Charts: Charting the Complex Relationships among Energy, Water, and Carbon," 2020.

even got to speak to A. J. Simon and the group who does this work, after I passed the comprehensive security checks associated with visiting LLNL. The 2019 and 1973 charts basically look the same: the same primary energy sources, the same economic sectors, and roughly the same ratio of useful to rejected (or waste) energy. If anything, it appears we waste more energy today, but that is more of an artifact of the subtly changing methodologies used in creating the chart than anything else.

The thinking that shaped the 1970s response has left us with a group of people who believe that energy problems can be solved with efficiency on the demand side (CAFE standards and Energy Star appliances), and those who think transformation is about creating more supply (whether it be nuclear or natural gas). This has mired us in an old way of thinking that constrains us from seeing the big picture today and the (I wish it were more obvious) fact that we must transform supply and demand simultaneously.

## SUPPLY MUST EQUAL DEMAND

Another consequence of the 1970s energy response is that it baked in a "supply-side" view of energy. When it was first conceived, the Sankey chart flowed from barrels of oil and tons of coal on the supply side to four big, opaque buckets—Industrial, Residential, Transportation, and Commercial—on the demand side. This gave us a perspective weighted toward a supply of precious rocks and magical energy-dense liquids, but not much insight into demand-side energy use. The thinking was that if America was more efficient with regard to these big economic segments, we could reduce the supply. The chart was capable of pointing in the general direction of which efficiencies would negate the need for imported oil: better mileage and more efficient homes and appliances (remember that many homes at the time were heated with oil, and some still are today). But it didn't give us much insight beyond that.

The federal agencies that Nixon set in motion and Carter brought into existence had also begun to collect a much richer set of data on the demand side of the picture. We now have high-resolution data thanks to semi-annual surveys of the industrial,[3] residential,[4] commercial,[5] and transportation sectors.[6] To this day, when people think about energy and transforming the energy system, they make the reasonable assumption that we will need the same amount of energy on the supply side as we've always needed. But, as we'll see, using all of the extra information on the demand side that these data sets give us, that's not the case.

The new methodology looks at all the things we like to do as human beings (our demands), and all the energy those needs require. We can then imagine how to decarbonize those needs and estimate the new amount of energy required to supply them—and, critically, which zero-carbon energy source to use (e.g., electricity or biofuels). These calculations quickly lead to the conclusion that we should electrify almost everything, and because electrical machines are inherently more effective, that we'll need far less energy on the supply side than you might think. There's no free lunch here, there's just a better lunch that we haven't been eating.

We should pause for a moment to give thanks to fossil fuels. When people started burning coal in quantity instead of biomass, from the mid-1700s to the mid-1800s, it kicked off the Industrial Revolution that freed humans from a lot of back-breaking labor. Fossil fuels were used to heat our homes, light our streets, power our railroads and steamships, refrigerate our food, and allow us quick and easy transportation in the form of cars, trains, and motorcycles. Coal, oil, and natural gas now power our modern lives, and for many of us, our modern lives are pretty darned good. Fossil fuels have been amazing.

But fossil fuels are now obsolete, because of the carbon dioxide they produce. We've come to the point where we have to substitute new energy sources for all our fossil fuels. We need to understand as much as we can about our demands for energy use, and then figure out how we are going to meet those demands, in detail. I have long been obsessed with energy data. I once measured—to my wife's chagrin—every single use of energy in my life, including the fuel in our cars and the electricity and natural gas in our home. I even weighed every object we owned so I knew what portion of my energy consumption our newspaper subscription and book collections represented. This resulted in me recommending that my wife cancel her daily newspaper subscription, as the 10 pounds of paper coming into our house each week represented a substantial component of our energy use. Avoiding divorce, we settled on a subscription to the Sunday paper!

After a decade of privately obsessing about energy data, in 2018 my company, Otherlab, was contracted by the Department of Energy (through ARPA-E) to take a closer look at all of the data that we have about our energy uses.[7] We pulled in data from the Residential Energy Consumption Survey (RECS), the Commercial Buildings Energy Consumption Survey (CBECS), the National Household Transport Survey (NHTS), the Transportation Energy Data Book (TEDB), the Federal Energy Management Program (FEMP), and the North American Industry Classification System (NAICS).

My job became one of reading the footnotes and following the trails to their end. We were tasked with building a tool that could help prioritize federal energy research and development spending. Naturally (for me) that ended up being summarized as a Sankey flow diagram of energy from its mining, production, and import, to its end uses in homes, factories, and even churches. I dragged a bunch of my colleagues, including Keith Pasko, Sam Calisch, Arjun Bhargava, Pete Lynn, and James McBride, through this obsession with me. We even had the final diagram printed on a shower curtain that hung in the office bathroom, and we still have giant versions of this poster on the walls of our office. We had loosely set ourselves the goal of tracking US energy consumption down to a tenth of one percent. We followed every energy flow and data set to its very end to see what we could learn. We went down every rabbit hole. We were able to achieve our goal, but the end result deserves the nickname given to Sankeys: "spaghetti charts."

Once you have all that information about our energy system, another thing you can do with it is start to think about how we might transform it. By visualizing the data, the need to transform the demand side at the same time as the supply side becomes clear. It also becomes obvious that not only is electrification the pathway to transformation, it comes with more built-in efficiency than we have ever experienced before. The complicated picture (shown in figure 4.8) that emerged is very detailed and utterly fascinating. I now bore dinner guests and public audiences with statistics about the proportion of all US energy used to drive children to church or school (0.7%); the amount of energy used in empty buildings (0.03%); the energy used to transport meat, fish, poultry, and seafood (0.2%); the amount of energy used in jet fuel for our military planes (0.5%); energy used in mobile homes (0.5%); energy used piping natural gas around the country across 4.4 million miles of pipeline (0.87%); and even the 0.005% of energy we use to light our billboards.

A huge amount of insight, and (at least for me) quite a lot of enjoyment can come from diving into the data to look at each sector of the economy and some of the unusual, or at least not obvious, ways that the US uses energy. It should be emphasized that this data summarizes all of the manifestations of our society in terms of its energy use and gives us quite a view into our collective human desires. Presented with all the numbers it is tempting to judge one energy use against another—such as the worthiness of the 0.24 quads used for recreational boating versus the 0.48 quads used in public assembly buildings—yet one should be wary of comparing these apples and oranges and applying moral judgment. Or, as my late friend David J. C. MacKay used to say, "All human activity is folly."

## THE GOVERNMENT SECTOR

Since 1975, the Federal Energy Management Program (FEMP) has monitored the energy use of US government agencies, as figure 4.3 shows. It provides a fascinating, if incomplete, picture of the energy uses our tax dollars support. The reason it is incomplete is that it only monitors things that are easily measured—petroleum, electricity, and natural gas use. It does not include the energy that went into building government office buildings, aircraft carriers, tanks and guns (all of which are filed under the "industrial" category). One initially surprising observation is that government energy use is completely dominated by the jet fuel used to support military operations around the world—nearly one half of one percent of all of our energy use! In distant second place in terms of our government's energy use is the US Postal Service, a fabulous institution, delivering the mail, rain, hail, or shine. NASA, comparatively, uses a tiny amount of energy to support the exploration of the universe and encourage our species to look toward the stars.

## THE RESIDENTIAL SECTOR

The next category is the residential sector, the one we are most familiar with. We see that the pride of the suburbs, the single-family detached home, dominates energy use, with large apartments in a distant second

# US Government Sector Breakout
## 1.3 Quads

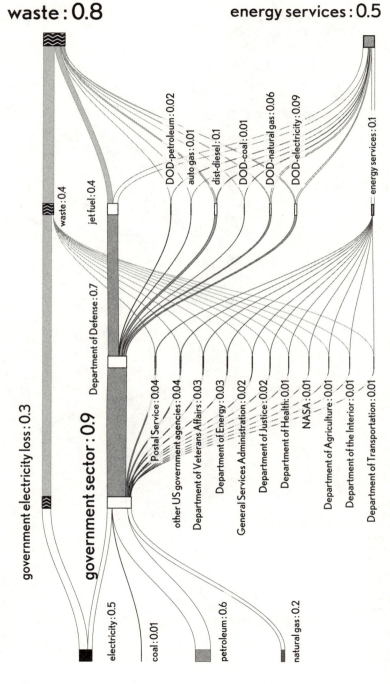

4.3   Energy flows from supply to demand in the US government sector.

# US Residential Sector Breakout
## 19.0 Quads

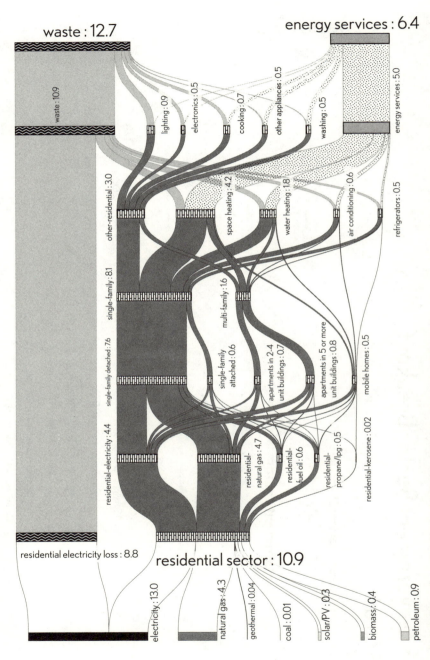

**4.4**  Energy flows from supply to demand in the US residential sector.

place. Roughly half of the total residential energy use goes to heating the spaces we live in. A quarter is attributed to water heating, while the final quarter is spent lighting, cooking, washing, and powering our electronic devices. When addressing climate change, we have to acknowledge the need to solve for all living situations. Mobile homes, for example, are an important, if neglected, component of our building stock. There are many efficiencies to living in homes that size, as the "tiny house" movement has underscored. Also vital when contemplating decarbonizing this sector is that we simply don't have time to build new versions of all of our homes. We won't solve climate change in time unless we figure out how to retrofit our current homes and dwellings for the electrified future we will live in. There are 130 million households in the US, around 95 million of them single-family homes, yet only 1.5 million homes are built each year—it would take more than 100 years to turn over the building stock.

## THE INDUSTRIAL SECTOR

The industrial sector is the largest energy consumer of all the economic sectors when we account for its (thermoelectric) losses. It's a complicated sector to unpack in terms of energy use. A massive amount of energy in this sector is used in finding, mining, and refining fossil fuels. It takes a lot of energy to pull coal, oil, and natural gas out of the ground and to convert it into refined products. We need significant amounts of natural gas to make plastics and fertilizers. This sector also includes a huge amount of biofuels. The paper pulp industry requires a lot of trees to produce paper, cardboard, newsprint, and building materials, and the byproduct is a biofuel that powers that industry. Looking at this sector reveals several surprising details that make fun pillow talk—though my wife doesn't always think so—like the fact that around 0.28% of energy is used in tending our major field crops, while another roughly 0.5% is used to grind and crush rocks.

Some segments of the industrial sector are so new that they are hard to get good data on. Data centers—the places where great swaths of the internet are stored—are one of those segments. Data centers are estimated

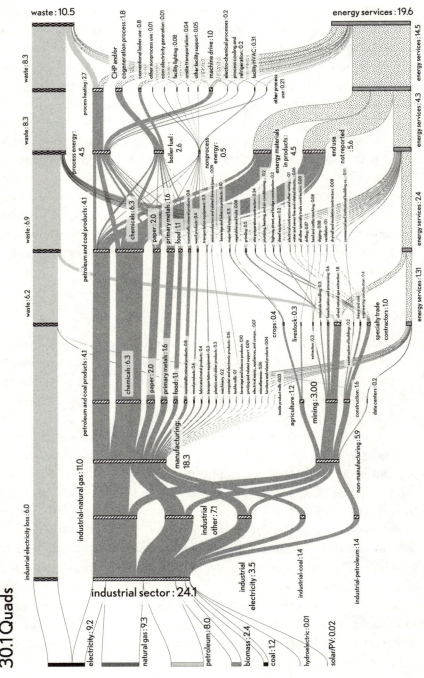

**4.5** Energy flows from supply to demand in the US industrial sector.

to account for about 0.25% of energy use (1% of electricity) today, and it is growing, but not as fast as some people feared.[8] A friend and one of my graduate school colleagues, Jason Taylor, runs infrastructure for Facebook. Periodically, we discuss energy use, because for data companies like Google and Facebook, energy use is critical to operations and often the biggest expense after payroll. While talking about how one would go about decarbonizing Facebook operations, Jason admitted, "We now have to work with a data paradigm of write once, read never." In other words, that photo of your kids you uploaded for Grandma will be seen only once, but it will require tiny amounts of energy forever as it is stored in some backwater memory bank. As with our disposable material culture, so goes our information, and because of that an ever-increasing amount of energy is required to keep all of our old data in the online shadows. If it's not happening already, I expect activists to call for data cleaning and recycling and "sustainable social media" in the near future.

## THE TRANSPORTATION SECTOR

The transportation sector is a close second to industry in terms of energy use. While air travel gets a bad rap, it is transport on highways that by far dominates this sector's energy use, using more than 10 times the energy of air travel. Of this highway energy, about 75% is expended by small vehicles, the passenger cars and trucks used to move ourselves around. Amazingly, almost half of this is used on trips of less than 20 miles, mostly to get to and from work and for family responsibilities—things like church, shopping, and school. Of non-highway transport, air travel is the largest contributor, followed by ships and then trains. Incidentally, a fully loaded modern jet aircraft gets the equivalent of around 60 miles per gallon (MPG) per passenger, so for traveling long distances, they beat solo road trips in cars (but if you take four friends with you, even a gas-guzzling American car is not so bad—something hyped by the ride-share community). We can even see that the energy required to transport fossil fuels is significant, with about 1% of US energy use committed to transporting natural gas (we'll come back to this later). Nearly half of freight-rail transportation is used to move coal—most of the other half is wheat and food. A not-so-surprising revelation from a close study of the

# US Transportation Sector Breakout
## 46.4 Quads

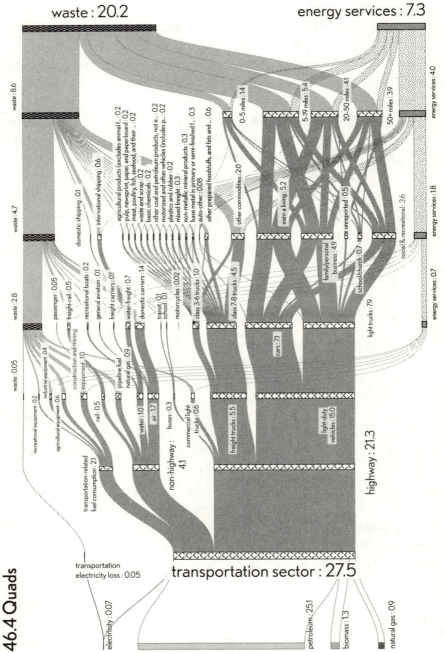

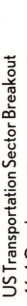

**4.6** Energy flows from supply to demand in the US transportation sector.

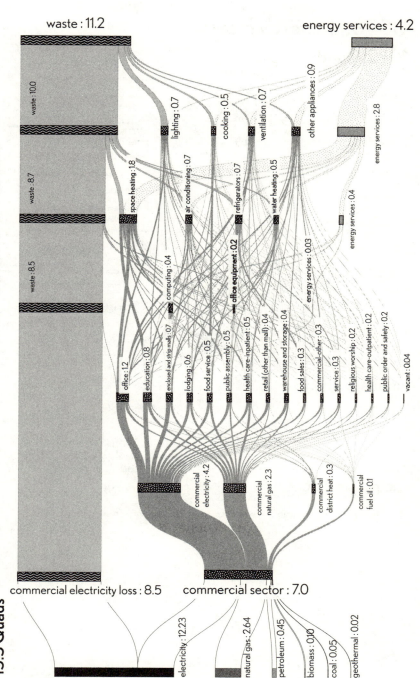

**US Commercial Sector Breakout**

**15.5 Quads**

**4.7** Energy flows from supply to demand in the US commercial sector.

# US Total Energy Flows
## 101.2 Quads

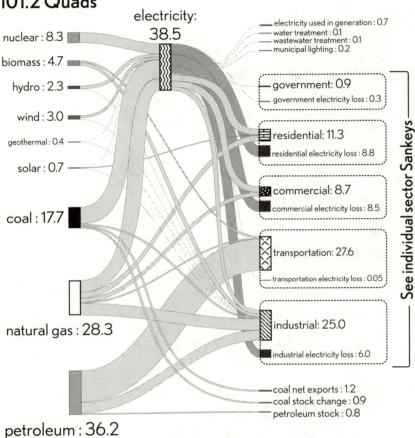

electricity: 38.5

nuclear : 8.3
biomass : 4.7
hydro : 2.3
wind : 3.0
geothermal : 0.4
solar : 0.7

coal : 17.7

natural gas : 28.3

petroleum : 36.2

electricity used in generation : 0.7
water treatment : 0.1
wastewater treatment : 0.1
municipal lighting : 0.2

government: 0.9
government electricity loss : 0.3

residential: 11.3
residential electricity loss : 8.8

commercial: 8.7
commercial electricity loss : 8.5

transportation: 27.6
transportation electricity loss : 0.05

industrial: 25.0
industrial electricity loss : 6.0

coal net exports : 1.2
coal stock change : 0.9
petroleum stock : 0.8

See individual sector Sankeys

**4.8** Energy flows through the whole US energy economy, from supply to sector.

Sankey diagram: our fossil-fuel supply chain is itself a major consumer of fossil fuels.

## THE COMMERCIAL SECTOR

The commercial sector encompasses all economic activities not associated with manufacturing or transportation. These activities are diverse, but the largest portion is used in office buildings and schools, mostly for space heating, water heating, and air conditioning. Hotels, malls, and hospitals are just behind, combining for about 20% of the total. The "cold chain" responsible for refrigerating our perishables on their way to our homes uses close to 10% of all commercial energy. The thermoelectric losses associated with producing electricity-using fossil fuels is a dominating factor in this sector today.

## THE BIG PICTURE

The complete dataset, all assembled into one pretty picture, is so dense with information as to be barely readable in book format. I include it here for completeness more than for utility. All of the data can be seen and played with at www.departmentof.energy—yes, we had fun choosing that domain name. The complexity of the modern world is woven into the crisscrossing lines of our energy economy. Thanks to public-sector institutions that we built in response to an energy crisis many decades ago, we know more about our energy needs than any other economy in the world. The task is to look at every single one of these energy flows— even the small ones—and then ask, "How can we achieve the same result, but without producing $CO_2$ as a side effect?" As an entrepreneur in the energy space, I actually use this giant chart to outline the big economic opportunities in the coming decades, and strategies and technologies for building great businesses while solving our emissions problems.

As you'll see, for the most part we already know how to do everything we currently do, but much more efficiently, by using renewable and clean electricity. What began as a problem from the past is now an opportunity for the future.

# 5

## 2020s THINKING

- It's not the 1970s anymore, and we're not facing a '70s energy problem that can be solved with efficiency. We need transformation.
- '70s thinking focuses us on lots of small decisions and distracts us from the big picture.
- '70s thinking muddles thermodynamic efficiency with energy saved through behavior change.
- '70s thinking leads to a narrative of deprivation.
- '70s thinking was about doing less bad, not about doing more good and building good into the way we do everything.

The United States is stuck in a way of thinking about the environment that dates back to the 1970s. This mindset can be succinctly summarized as (pardon my Australian), "If we try extremely hard, and make many sacrifices, the future will be a little less fucked than it might be otherwise."

To address climate change, we need a new narrative that is both more honest about the task at hand and more broadly engaging than a story about sacrifice. It can be a story about what we stand to win—a cleaner electrified future, with comfortable homes and zippy cars—which is better than nightmares about what we have to lose. We have a path to decarbonization that will require changes, to be sure, but not deprivation. The

2020s mindset says, "If we build the right infrastructure, right away, the future will be awesome!"

The language of sacrifice associated with being "green" is a legacy of 1970s thinking, which was focused on efficiency and conservation. The 1970s began with Earth Day (April 22, 1970), and was a decade defined by two energy crises over oil imports. The air[1] and water quality[2] problems caused by our energy production were coming to the fore, in part because of groundbreaking books like Rachel Carson's *Silent Spring* and the burgeoning environmentalist movement they inspired. The answer to these problems became a story about conservation: use less fossil fuel, turn down the thermostat, buy smaller cars, and drive less. This is the era that gave us the mantra "Reduce! Reuse! Recycle!"

This approach translated to more fuel-efficient (but still petroleum-burning) cars and better-insulated homes (but still heated with natural gas). The emphasis on efficiency ever since the '70s is reasonable, since almost no one can defend outright waste, and almost everyone agrees that recycling, double-glazed windows, more aerodynamic cars, more insulation in our walls, and industrial efficiency will make things better. But while efficiency measures have slowed the growth rate of our energy consumption, they haven't changed the composition. We need zero-carbon emissions, and, as I often say, you can't "efficiency" your way to zero.

The '70s emphasis on efficiency was also confusing, in that it conflated different types of efficiency. You can make a big car more efficient with a more efficient engine, or you can buy a smaller car that is more efficient because of its smallness, or you can use your car less. The first of these efficiencies is thermodynamic efficiency; the other two come from behavior changes. Environmentalists have focused more on behavior-change efficiencies—which are fine, as far as they go—but we will gain a lot more with big technological changes. Rather than make a more efficient fossil fuel-powered car (thermodynamic efficiency), or drive it less (behavior efficiency), it makes more sense to make an electric car powered by renewable energy.

2020s thinking is not about efficiency; it's about transformation.

Nearly half a century after President Jimmy Carter's famous remarks about energy conservation measures, turning down the White House thermostat, and wearing a cardigan to prove the point, experts know efficiency fixes are not enough. Often forgotten is that Carter's comments were oddly similar to what Nixon said six years before him.[3] While we have made more fuel-efficient appliances and paid a lot of attention to "greening" our small daily purchases, we haven't done much to solve our larger carbon problem. And even if energy efficiency were sufficient, America hasn't shown any inclination to drastically cut its consumption since the '70s.

Also, Americans will never fully support decarbonization if they believe it will lead to widespread deprivation—which many people associate with efficiency. We can't address climate change if people remain fixated on, and fight about, losing their big cars, hamburgers, and the comforts of home. A lot of Americans won't agree to anything if they believe it will make them uncomfortable or take away their stuff.

The environmental movement needs to stop focusing on efficiency— and on the demand side of the energy equation in general, which says that if we just use less, we will need to supply less. Nor can we simply address climate change by greening the supply unless we also swap out all our machines on the demand side. We need an entirely new paradigm, which isn't mired in our '70s notions of supply and demand, but realizes that the two are intimately connected. America needs to decarbonize supply at the same rate as it decarbonizes demand, and that means powering electric machines with zero-carbon electricity.

It's 50 years later. Now, we must play end-game decarbonization.

In this 2020s paradigm, environmentalists need to think bigger. We need to change our mindset from the efficiency environmentalism of the 1970s to a transformation mindset appropriate to the twenty-first century. Efficiency proponents may counter that if you make the thing more efficient first, then you will need less electricity. That is true. But I would then argue that electrification is more politically palatable and offers a bigger immediate win, and that we should look at this problem from all sides to make the most progress. Let's stop imagining that we can buy enough sustainably harvested fish, use enough public transportation, and purchase enough stainless steel water bottles to improve the climate situation. Let's release ourselves from purchasing paralysis and constant guilt at every small decision we make so that we can make the big decisions well.

Instead of efficiency, massive electrification is the best way to address climate change. If the electric car in your driveway is powered by the solar on your roof and in your community, and your heating system runs on electricity generated at a far-off wind farm, then you have already made the small number of critical decisions that eliminate the large majority of emissions from your life.

End-game decarbonization means electrifying everything. It means that instead of changing our energy supply or demand, we need to transform our infrastructure—both individually and collectively—rather than our habits.

# 6

## ELECTRIFY!

---

- America can't replace fossil fuels with similar fuels just because they feel familiar.
- The country can't keep burning fossil fuels and assume we can suck $CO_2$ out of the air and stuff it back into the earth or oceans.
- We have to electrify (nearly) everything.
- When we electrify everything, we find we'll use about half the energy we currently use.

We can't use fossil fuels. So how will we keep the world running?

When people imagine switching to zero-carbon energy, they often think about simply swapping fossil fuels for another "familiar fuel." If you had a gallon-sized gas container, you'd like to just fill it with something else that is zero-carbon yet powers the same lawnmower, or a familiar-looking car. That's why people think a lot about net-zero carbon fuels. Biomass, ethanol, switchgrass, sargassum—there are lots of things that absorb $CO_2$ from the atmosphere as they grow and emit it when they burn. Couldn't these fuels be used in our machinery with minimal changes to life? It sure sounds like it could work.

Similarly, people talk about producing hydrogen or synthetic fuels like ammonia or ethanol with properties similar to gasoline or natural gas.

Again, it sounds easy, but it requires using more renewably or nuclear-generated electricity to create the fuels than you would need to simply power an electric car straight from the grid. Hydrogen vehicles are the canonical case of this silliness. The idea is to make a unit of electricity, lose 25% of it in converting it to hydrogen, and lose another 25% of it in a fuel cell that converts it back into electricity that powers the wheels—all for the convenience of having a familiar fuel to fill a familiar tank. Nearly all hydrogen now used in hydrogen vehicles is a byproduct of natural gas, which just perpetuates our current problem, and is part of the reason these fuels have been cynically over-promoted as a solution.

The efficiency of a total-energy pathway is the sum of the efficiencies of the component pathways. To illustrate the point, let's look at three ways to power a car: via electricity, hydrogen, or some magical gasoline-like fuel produced from electricity (the latest entrant in this shell game is Prometheus Fuels, complete with advertising copy that makes you believe you can save the world with your old Ford Mustang).

In an electric car, we take the electricity, store it in a battery (~90% efficient), and then pass that electricity through a drivetrain that is about (~80% efficient).

$$Total\ efficiency = 1\times0.9\times0.8 = 0.72 \tag{6.1}$$

We get 0.72 units of transportation for one unit of electricity.

If we use the same electricity to make hydrogen (via electrolysis, ~65% efficient), then compress it into a tank and decompress it back out (~75% efficient), then run it through a fuel cell (~50%):

$$Total\ efficiency = 1\times0.65\times0.75\times0.5 = 0.24 \tag{6.2}$$

We get only 0.24 units of transportation for the same one unit of electricity.

If we had a process to make gasoline from electricity, it would probably be ~50% efficient, and the gasoline would only move the car at about ~20% efficiency, which looks something like:

$$Total\ efficiency = 1\times0.5\times0.2 = 0.1 \tag{6.3}$$

Or a mere 0.1 units of transportation for the same unit of electricity.

Given how hard it is going to be to create all of the electricity we need (as we'll see in the next chapter), it is very difficult to believe that we'll make three or even five times as much just for the convenience of having a fuel that was familiar in the twentieth century. It would be as if Henry Ford tried to make gas-powered metal horses.

Some version of this basic math applies to all of our decarbonization choices.

The biofuel route imagines it's possible to make a similar amount of fuels with biomass, but the problem is that there just isn't enough to go around. To create the amount of biofuel the world needs to power itself, we'd have to burn a quarter of all the biomass that grows on earth each year, every year, which would have devastating environmental consequences. At best, we can make about 10% of our fuels this way.[1]

The synthetic-fuel route imagines we make carbon-free electricity using solar, nuclear, wind, and hydroelectricity, and use that electricity to make the molecules of fuels similar to those we currently use. This is a game of compounding inefficiencies, as we saw above.

Imagining that we can just swap one fuel for another will keep the US stuck in the 1970s world of lots of machines with low thermodynamic efficiencies. It also keeps the country tied to the massive inefficiency of burning enormous amounts of material. As I explore in chapter 17, humanity moves more tons of fossil fuels than any of the other things that humanity produces—more than all of our agricultural products, more than all of our metals and ores. Imagining that we will build an industry that can manufacture this amount of alternative fuel on the necessary timeline is absurd.

The other "familiar fuel" strategy is carbon sequestration, whose proponents imagine that we'll use the same fossil fuels, suck the $CO_2$ out of the atmosphere, and bury it. Again, the tons of $CO_2$ humanity produces every year is more than all of the other material flows we use *in total*. There is no dump large enough to bury those emissions, even if it wasn't a thermodynamically awful idea in the first place.

What do I mean by a thermodynamically awful idea? Carbon sequestration requires the use of even more fossil-fuel energy (about 20% more) just to capture the $CO_2$ produced by those fuels, then uses yet more energy to compress it and bury it—and hope that it stays buried, something that isn't guaranteed.

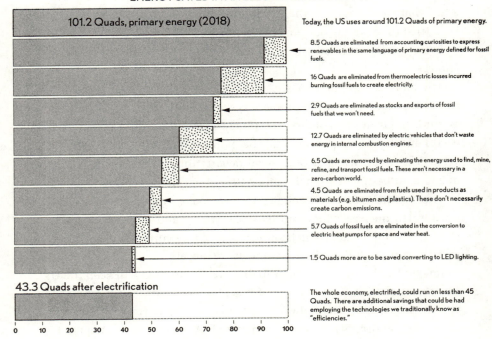

ENERGY SAVED IN AN ELECTRIFIED ECONOMY

101.2 Quads, primary energy (2018)

Today, the US uses around 101.2 Quads of primary energy.

8.5 Quads are eliminated from accounting curiosities to express renewables in the same language of primary energy defined for fossil fuels.

16 Quads are eliminated from thermoelectric losses incurred burning fossil fuels to create electricity.

2.9 Quads are eliminated as stocks and exports of fossil fuels that we won't need.

12.7 Quads are eliminated by electric vehicles that don't waste energy in internal combustion engines.

6.5 Quads are removed by eliminating the energy used to find, mine, refine, and transport fossil fuels. These aren't necessary in a zero-carbon world.

4.5 Quads are eliminated from fuels used in products as materials (e.g. bitumen and plastics). These don't necessarily create carbon emissions.

5.7 Quads of fossil fuels are eliminated in the conversion to electric heat pumps for space and water heat.

1.5 Quads more are to be saved converting to LED lighting.

43.3 Quads after electrification

The whole economy, electrified, could run on less than 45 Quads. There are additional savings that could be had employing the technologies we traditionally know as "efficiencies."

**6.1** Massive electrification scenario of the US energy economy that models primary energy reductions by sector and end use. Electrification of the economy with zero-carbon sources reduces our energy needs by more than half. The big wins are in eliminating waste heat from electricity generation, in the far greater efficiency of electric vehicles and electrified heating systems, and in eliminating the significant amounts of energy used in finding, mining, refining, and transporting fossil fuels.

Given that renewable electricity is already cost-competitive with fossil fuels, it is fairly obvious to those who think about energy that the cost of carbon sequestration makes it economically unfeasible.

All of these ideas are cynically promoted by people who wish to keep profiting from fossil fuels, burning your children's future. Don't let them divide us by confusing us. We don't just need to change our fuels; we need to change our machines. We need to use 2020s thinking to reimagine our infrastructure.

At the highest level, any realistic plan for total decarbonization is simple: **electrify everything**.

We have the technology we need, today, to solve climate change. And when we electrify everything, as I'll soon show, we will cut our energy needs in half!

## WHERE DOES ALL OF OUR ENERGY GO?

If you sit down with all of the data on the total amount of energy we use in the US, and begin with the thought experiment "what happens if we electrify everything?" some interesting things jump out. This is what I have illustrated in figure 6.1, where you can see that we need less than half of the primary energy that we think we do, which makes the task of generating it with renewables twice as easy. Here's how.

### MAKE CLEAN ELECTRICITY, SAVE ~23%

We can eliminate almost a quarter of the energy we think we need if we stop burning fossil fuels to generate electricity.

In a power plant today, fossil fuels are burned to generate heat, which is used to make steam, which is used to spin a turbine, which is used to create electricity. Physics tells us that using heat to generate electricity is subject to inescapable limits on efficiency. Those limits are set by the laws of thermodynamics, which dictate that machines that convert heat to electricity lose half or more of the energy involved in the conversion. This is known as Carnot efficiency—named for Nicolas Leonard Sadi Carnot, the man often referred to as "the father of thermodynamics"—which is the ratio of ambient temperature to the temperature of combustion. Under most real-world circumstances, fossil fuel–burning machines are 20–60% efficient.

Carbon-free, non-thermal sources like solar and wind—while also subject to the laws of physics—don't involve as many conversions from one type of energy to another. Because of this, generating electricity with renewables would eliminate approximately 15% of the primary fossil energy we currently think we need to run the economy. Today, when we use fossil fuels we are harnessing a long train of energy conversions: solar energy from long ago was converted to a biofuel (a plant or dinosaur), which over geological time became a fossil fuel, which was burned to become heat, which evaporated water to become steam, which spun a turbine to become motion, which through electromagnetism became electricity. These processes all wasted a little or a lot of energy at each step along the way. When we use solar panels, a photon from the sun strikes a semiconductor and liberates an electron due to the photoelectric effect (thanks, Einstein). So, although solar is typically only 20% efficient, we aren't losing hard-won energy the way we do in a 20%-efficient car engine.

Other "savings" in the amount of energy we think we need come from a couple of longstanding accounting curiosities associated with fossil fuels, which have caused us to overestimate the primary energy required to produce both hydroelectricity and nuclear energy. Primary energy is a useful, if imperfect, measure of the amount of energy needed to run the country. Traditionally it was the measurement of the tons of coal, cubic feet of natural gas, and barrels of oil that were the primary inputs into the economy. With nuclear and renewable energy as new energy options, the question of what is primary energy became important.

In the 1970s, concerns about scarcity and drought led scientists to calculate the primary energy of hydroelectricity to be the amount of fossil-fueled power that would need to be added to the grid to replace a hydro-facility lost to a drought. Because the average efficiency of fossil fuel–fired electricity generation is only around 30–40%, this resulted in an overestimate of the primary hydroelectric resource, which has persisted in accounting practices to date. Strangely, in calculating hydro-energy, we take the capacity of the hydro-facility and divide it by the average inefficiency of our fossil fleet—which means we over-report by a factor of three. Such are the oddities of a world literally defined by fossil fuels.

The second accounting curiosity is how we measure primary energy in nuclear-generated electricity. The US elected to use light-water reactors for its nuclear electricity plants, in part because of security and safety issues associated with the resulting waste. In this type of reactor, only about 1–2% of the energy in the fissile material is extracted; however, this process avoids dangerous or weaponizable isotopes in the reaction chain. We could have used breeder reactors like France and Germany do, which generate more fissile material than consumed, although these reactors create an even more difficult set of safety and security issues. Instead of using the conversion efficiency of nuclear fuel to useful energy to measure the efficiency of nuclear plants, the US Department of Energy decided to use the "heat rate." This, in effect, is just the thermodynamic efficiency of the steam turbine at the output end of the power plant, and it ignores what happens in the reactor. In the context of figuring out how to decarbonize America, choosing heat rate to define the efficiency of a nuclear plant ignores the other 98% of the fuel we don't use and simply doesn't reflect the efficiency of nuclear.

These accounting errors can mislead people into thinking there is more waste in the energy system than there really is, and they ignore the other technological options for harnessing nuclear energy.

If we correct these accounting errors mired in fossil-fuel thinking, we see that roughly 8% of the energy we think we need was never really there. Together, the savings from thermodynamic efficiency and proper accounting total around 23%, just for switching electricity generation to carbon-free sources and choosing a twenty-first-century accounting methodology. This all sounds complicated, but at the very least the upshot is we need less energy than we think to supply a fossil-free world; solving climate change is 10% easier to solve than you thought before you read this.

## ELECTRIFY TRANSPORTATION, SAVE 15%

Electrifying transportation is the next big energy win, saving us around 15%. Gasoline car engines, which make up the overwhelming majority of vehicles today, are even less efficient than power plants in converting fossil fuels into useful energy. By the time the energy in the fuel has been turned into the motion of the vehicle, the efficiency is only about 20%. The heat you feel on the hood of your car after a long drive represents some of that waste. (You can fry an egg on your engine block—with old Land Rovers, people learned to put a Dutch oven in the engine bay and cook a stew as they drove!) By electrifying all cars and trucks, we will eliminate most of that waste heat and cut the amount of energy consumed in powering those vehicles by a factor of three.

Electric cars have gone mainstream. They are becoming more affordable and faster charging, and they are expanding in performance, range, and options. At their current rate of improvement, we are only a few years away from electric vehicles with a range of 500 miles. We already have vehicles with enough range for nearly all purposes except for extreme road trips or extraordinarily long days. It is not a matter of if, but when.

## ELIMINATE FINDING, MINING, AND REFINING FOSSIL FUELS, SAVE 11%

A huge and largely unseen amount of energy is used to discover, mine, refine, and transport fossil fuels. In a zero-carbon economy, we won't need to expend that energy, which saves us more than 11%. Oil and gas extraction consume nearly 2% of US energy flow. Transporting natural gas (1%), running coal-mining equipment (0.25%), moving the coal from the mine to the power plant via rail (0.25%), and refining crude oil into gasoline and diesel (3–4%) together consume around 8% of the national energy supply. It surprised me to learn that nearly half of all of the tonnage of stuff that is moved by rail is coal; roughly the other half is grain, along with a few cars and pieces of machinery, and a small number of people. Our accounting isn't exact, because we have to consider the amount of coal, natural gas, and oils in stockpiles—strategic reserves—which varies, but our total savings is about 11%. In all likelihood, the energy savings of eliminating fossil fuels is even greater, especially if we account for the fuel for tanker trucks delivering gasoline from refineries to gas stations and the energy that goes into building all of the mining and shipping equipment necessary for this massive heavy industry. Remember, the US moves as many tons of fossil fuel as any other commodity category (I'll explore this in detail in chapter 17).

The thoughtful reader might argue that these savings will be offset by the energy required to build the windmills, solar cells, batteries, nuclear plants, grid, and electric vehicles that will replace the fossil fuel industry. But the energy used in their construction and operation is likely a significantly smaller percentage of the future energy economy than fossil-fuel processing is today. Energy returned on energy invested (with the world's worst acronym, EROI) describes how much energy you have to put in to get some amount of energy out. We just saw that the EROI of fossil fuels is around 7–8. One unit of fossil fuel energy in gets you 7 or 8 units back. Historically, the EROI of fossil fuels has not accounted for inefficiencies in electricity generation, making them look more efficient than they really are. When this is taken into account, renewables beat fossil fuels, hands down.[2] Estimates may vary, but wind and solar provide approximately twice the EROI of fossil fuel power plants. As manufacturers reduce the energy input of producing wind and solar technology, and as engineers extend the useful lifetime of this green machinery, the advantage will only improve.

### ELECTRIFY BUILDINGS, SAVE 6–9%

Electrifying the heat used in homes and offices offers another huge opportunity for savings in the new energy economy. For low-temperature heat (e.g., thermal energy that is hotter than human skin but cooler than boiling water), we have an astounding and well-developed technology called heat pumps that significantly outperform the old ways of doing things.

Today, space heat and hot water for homes and offices is usually created by burning natural gas or fuel oil, or by running electricity through a resistive heating element. Heat pumps work on a different principle, concentrating thermal energy from an abundant source, such as the air outside or earth underneath your house, into household appliances and HVAC equipment. This difference allows them to operate more efficiently, supplying more than three times as much heating or cooling per unit of energy input as conventional approaches. If deployed at scale in the US, these devices would cut another 5–7% of the total energy the country requires.

LED lighting wins us another 1–2%. Like electric cars, LED technology has matured greatly in quality, performance, and availability in the past few years. Lumen for lumen, LEDs use one-fifth the energy of traditional lighting technologies. What's more, they last for tens of thousands of hours and require much less frequent bulb replacement. Integrated controls and occupancy sensors for switching off lights when they aren't needed can extend these savings further. A wholesale commitment to these technologies can save us another 1–2%.

## ACCOUNTING FOR FOSSIL FUELS WE DON'T BURN, SAVE 4–5%

Fossil fuels that get turned into our day-to-day materials currently account for 4–5% of our "energy use." Rather than being burned to provide power, they are transformed into familiar products. A common example is black-top roads, which are partly made out of bitumen (asphalt), a byproduct of the oil-refining process. Bitumen is also a component of 85% of the rooftops (asphalt shingles) in America. Plastics are made using material derived from natural gas. Carbon from coal is used to turn soft iron into carbon steel. Most of the carbon in these materials isn't released into the atmosphere as $CO_2$, so their energy content is not relevant to today's climate conversation. While we should track their use, it should be in a resource assessment of material flows and sustainability constraints, not in terms of their impact on the energy economy.

## ELECTRIFY MANUFACTURING

Huge energy savings are possible by electrifying industry, but we don't even need to account for that here to see the enormous benefits of elec-trifying the economy. I will cover the manufacturing sector and its con-tribution to environmental and climate success in more detail in chapter 17. In short, opportunities for innovation abound in this sector and make the energy-saving outlook for America even rosier.

## SAME COMFORT, SAME CONVENIENCES, HALF THE ENERGY

When we add up all of those savings, we find we only need ~42% of the primary energy we use today.

**Well, that is pretty remarkable.**

America can reduce its energy use by more than half by introducing no efficiency measures other than electrification. No thermostats were turned down, no vehicles were downsized, no homes were shrunk. Not only that, but electrification is a "no-regrets" option—we can also deploy other strategies like behavior change and the things we typically call efficiency, and see even further gains. That's why electrification is the only real strategy for decarbonization—and why it will release us from a paralysis of "what to do?" and provide us with an immune-response to those who seek to confuse the public with stories of the role of fossil fuels in our future.

There are too many people who quote too many numbers about the future with too much confidence. Yes, I can state that we probably only need 42% of the primary energy we need today, but that is overly granular, and future developments are likely to alter this figure. The population will grow slightly. We'll invent some cool new pastimes that use a bit more energy (electric-powered paragliding, anyone?). In the meantime, quality-of-life increases typically require an increase in energy consumption. Taking these variables into account, it is simplest to say that Americans will only need half the energy they use today if we electrify everything while improving our lives. What a win.

Winning the war against the climate crisis will also mean a cleaner, more positive future. Our houses will be more comfortable when we shift to heat pumps and radiant heating systems that can also store energy. While it may also be desirable to downsize our homes and cars, this isn't absolutely necessary, at least in the US. Our cars can be sportier when they are electric. Household air quality will improve, as will public health, since gas stoves raise the risk of asthma and respiratory illnesses. We don't need to switch to mass rail and public transit, nor mandate changing the settings on consumers' thermostats, nor ask all red meat–loving Americans to turn vegetarian. No one has to wear a Jimmy Carter sweater (but if you like cardigans, by all means wear one)! And if we sensibly employ biofuels, we don't have to ban flying.

In short, the climate-friendly future will be quite recognizable in terms of the major objects in our lives—our cars, homes, offices, furnaces, and refrigerators. All of these objects will just be electric. There is no need to fear this future, and there will be cost savings and health benefits if we embrace it—oh, and this will also address climate change at the same time.

# 7

# WHERE WILL WE GET ALL THAT ELECTRICITY?

---

> ☞ There are enough renewable resources to easily meet global energy demands.
> ☞ Solar and wind will be the biggest suppliers.
> ☞ Hydroelectricity is critical, especially as a giant battery.
> ☞ Biofuels matter, especially for things like air travel, but they won't solve every problem.
> ☞ Nuclear, while not strictly necessary, is very useful.
> ☞ Our land-use patterns are crucial to success.

To electrify everything, America will need more than three times the amount of electricity that it currently produces. Today, the US grid delivers, on average, 450 gigawatts (GW) of electricity. If we electrify nearly everything, as I described in the previous chapter, we'll need 1,500–1,800 GW. That's a lot. If we use solar alone, that's more than we can fit on all of our rooftops, and more than we can erect over our parking spaces (see figure 7.1). If we added wind turbines in all of the corn fields in America, that would supply about half of what we need. To arrive at this number, I assume a power density of wind of 2 W/m$^2$ based on standard turbine spacing in wind farms,[1] and a total acreage of corn, America's largest crop, of 90 million acres.[2] Of course, adding wind turbines and their supporting infrastructure can't be done without taking a small portion of the land from crop production, but this provides a sense of scale—and it also underscores the critical role of the agriculture industry and farmers in succeeding at decarbonization.

The good news is that there is no shortage of energy. The amount of solar radiation that makes it through our atmosphere and into our earth systems is 85,000 terawatts. A terrawatt (TW) is a trillion watts, or about the same power as one hundred billion LED lightbulbs. What this means is that the amount of solar that hits the earth far surpasses the approximately 19 TW that humanity uses.[3] The US uses approximately 20% of that, 3.5–4 TW of primary energy.

The sun is the primary source of almost all renewables, i.e., energy that can be replenished. The major player is solar, abundant wherever the sun shines. The sun heats the air and creates wind that can be harnessed with turbines. The winds whip up waves that can be captured by wave-power generators. The sun evaporates water, which becomes clouds and rain, filling rivers that can be tapped for hydroelectricity. As you know when walking on hot sand on a summer beach, the sun also heats the ground. This "ground-source" geothermal heat can be harvested year-round by a technology called heat pumps to keep buildings at an even temperature.

Confusingly also called "geothermal energy" is the energy that is a closer relative of geysers, volcanoes, and hot springs. These types of geothermal energy are not derived from solar, but from remnant heat left over from the formation of the earth, with a little heat generated from radioactive decay thrown in for good measure. This creates extremely hot rock, which is accessible by drilling, and can be used to create steam, which drives a turbine to create electricity. Horizontal drilling and associated fracking technologies can be appropriated to access more of this resource (in fact, the US possesses an amazing amount of this energy at a depth of 5–10 km), but this technology is still far from being proven cost effective.

The sun is also critical to photosynthesis, which creates biomass (wood, algae, grasses, forestry and agricultural waste, food waste, human waste, and other biological matter), which can be converted to biofuels to supply energy to difficult-to-decarbonize sectors like long-haul aviation. In fact, all of the world's fossil fuels are just very old biofuels that have been buried and concentrated over time.

## WHICH ENERGY SOURCES WILL WE USE?

Given America's energy needs, we'll have to make electricity wherever we can, while understanding that some sources are easier, cheaper, and more convenient than others. Some areas of the country have better wind, some have better solar, and some don't have enough of either and will probably require some nuclear. Where there are rivers, hydroelectricity, which provides nearly 7% of electricity in the US today, will be critical. Where there are oceans, wave and tidal power will help at the margins. Offshore wind is likely to be the big producer from the oceans.

Solar, wind, and nuclear are the resources at our disposal that far exceed US energy demands. Solar and wind are the cheapest, and have fewer complications than nuclear energy. There is so much money in the fight over the future of our energy supply that an enormous brouhaha emerged in the climate and energy world when Mark Jacobson of Stanford University,[4] along with his colleagues, proposed that the world could run 100% on water, wind and solar (WWS).[5] The pushback to this proposal was vicious,[6] and even by academia's petty standards, the rebuttal to the critique was even more vicious again,[7] with yet more venom in the rebuttal to the rebuttal.[8] It ended in a lawsuit. I believe history will side with Jacobson, and we'll be able to do this with WWS technology— and others agree with me.[9] Critics of the Jacobson plan argue that we can't have the reliability we need in an all renewables world. I tackle this issue head-on in the next chapter, and you'll see there is every reason to believe it's easier than we think to turn these intermittent sources into a reliable energy supply. We absolutely do need to think about supply and demand, and my critique of this academic storm in a teacup is that everyone involved in the argument should have paid more attention to both sides of the equation. Jacobson may be too anti-nuclear, but his critics are too anti-future.

We're blessed with enough zero-carbon energy to meet our needs and even expand our wants—we just have to harness that energy sensibly. Nuclear energy isn't renewable—there is a finite amount of fissile material in the world (primarily types of plutonium and uranium).[10] Our best estimates suggest we have 200–1,000 years of fissile material left, depending on what portion of the supply it will meet—and whether the US sticks with light-water reactors that don't produce weaponizable byproducts, or whether we move to breeder reactors that do. While the country could get by without nuclear energy, it is available to us and useful in places that don't have enough land area to support wind and solar infrastructure.

Regardless of the minutiae of how we decarbonize, solar and wind will do the heavy lifting. The no-regrets pathway to quickly transform our fossil fuel–powered world to a world powered mostly by electricity is a combination of a majority of renewables (solar, wind, hydro, geothermal) with moderate nuclear and some biofuels as a backstop.

The exact balance of those sources will vary geographically and can be determined largely by market forces and public opinion about land use. The details and balance of power (energy nerds are always good for a power pun!) will be determined by how well we can store electricity to address the variability of renewables, as I will discuss in chapter 8.

## HOW MUCH LAND WILL WE NEED TO USE?

The American landscape will necessarily look different when we make this switch to renewable energy. Solar panels and windmills will become pervasive in our cities, suburbs, and rural areas. To power all of America on solar, for example, would require about 1% of the land area dedicated to solar collection—about the same area we currently dedicate to roads or rooftops (see figure 7.1). Rooftops, parking spaces, and commercial and industrial buildings can do double duty as solar collectors. Similarly, we can farm wind on the same land used to farm crops.

As we've seen, to electrify everything in the US, we'll need to generate around 1,500–1,800 GW. To generate all of that with solar would take about 15 million acres of solar panels. You can check my numbers: I assume a real fill fraction (the percentage of ground covered by solar) of 60%, a cell efficiency (the amount of incident solar energy converted to electricity) of 21%, and a capacity factor (the effective percentage of the day that receives a full dose of sunlight) of 24%. So to get 1,500 to 1,800 GW we need 15 million acres, or roughly a megawatt per acre. To harness the same amount of energy with wind power alone would take around 100 million acres planted with wind turbines. For reference, the area of all US land is about 2.4 billion acres.

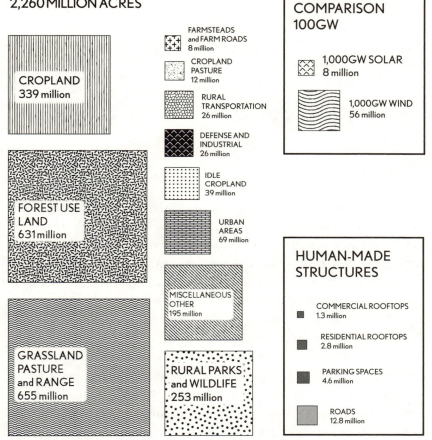

## LAND USE IN THE UNITED STATES
## 2,260 MILLION ACRES

CROPLAND
339 million

FARMSTEADS
and FARM ROADS
8 million

CROPLAND
PASTURE
12 million

RURAL
TRANSPORTATION
26 million

DEFENSE AND
INDUSTRIAL
26 million

IDLE
CROPLAND
39 million

URBAN
AREAS
69 million

FOREST USE
LAND
631 million

MISCELLANEOUS
OTHER
195 million

GRASSLAND
PASTURE
and RANGE
655 million

RURAL PARKS
and WILDLIFE
253 million

## RENEWABLES
## COMPARISON
## 100GW

1,000GW SOLAR
8 million

1,000GW WIND
56 million

## HUMAN-MADE
## STRUCTURES

COMMERCIAL ROOFTOPS
1.3 million

RESIDENTIAL ROOFTOPS
2.8 million

PARKING SPACES
4.6 million

ROADS
12.8 million

**7.1** Illustrative areas of the land use in the US, including reference areas for sufficient renewables to run the whole country. Offshore areas suitable for wind-power generation are not included.

**Table 7.1** Estimates of Land Area Occupied by the Country's 6 Million Commercial Buildings, 120 Million Homes, 8.8 Million Lane-Miles of Roads, and at Least 1 Billion(!) Parking Spaces

| Human-built thing | Million acres |
|---|---|
| Commercial rooftops | 1.2 |
| Residential rooftops | 2.8 |
| Roads | 12.8 |
| Parking spaces | 4.5 |

Some people talk about a solar cell we'll build in the center of the Arizona desert that will power all of America. But that's not actually how this job will get done, because of the expense of transmission (over long distances) and distribution (over short distances). Renewable-energy installations will be everywhere, so it is more illustrative to compare solar and wind generation to other ways humans use land. Because it requires a lot of land to power the country with solar and wind, it is worth looking at surfaces and activities that can do two jobs at once.

Let's first look at solar. In table 7.1, I present the acreage of all rooftops, roads, and parking spaces in the US—all places where we could install solar panels. Obviously, there are details about how to effectively use these land areas for renewable generation, but these figures are merely meant for comparison. For instance, paving roads with solar gets a lot of attention, but it isn't a great idea due to the dirt and abuse that results from driving cars on solar cells. It's better to think about putting solar panels in highway medians and lofting panels over roads and parking spots.

All of these add up to 21 million acres. If we were to use only solar to produce all our electricity needs, we would need nearly 15 million acres for panels—more than two-thirds of all available roofs, roads, and parking spaces. Clearly, we will need to put solar panels wherever we can fit them. There is a camp of environmentalists that believe we'll power the world with distributed (rooftop or community) solar, but the numbers tell a simple story that we'll need all of the distributed energy we can harness, and we'll need industrial installations of solar and wind as well.

Fortunately, we can also rely on the United States' abundant wind resources. Let's take a look at where we can put wind turbines. Again,

**Table 7.2** Major Land Uses

| Human land use | Million acres |
| --- | --- |
| Cropland used for crops | 339 |
| Idle cropland | 39 |
| Cropland pasture | 12 |
| Grassland pasture and range | 655 |
| Forest-use land | 631 |
| Rural transportation | 26 |
| Rural parks and wildlife | 253 |
| Defense and industrial | 126 |
| Farmsteads and farm roads | 8 |
| Urban areas | 69 |
| Miscellaneous other land | 195 |

Source: Daniel P. Bigelow and Allison Borchers, *Major Uses of Land in the United States, 2012*, EIB-178, US Department of Agriculture, Economic Research Service, August 2017.

they can do double duty, harnessing wind on agricultural and range-lands, among other places. Let's look at table 7.2, which shows land use in the United States.

Right away we can see that we have plenty of cropland, where we can also put wind turbines. Idle cropland is ideal for turbines (and perhaps for generating income for farmers). We also have massive amounts of grassland pasture and range that are similarly suitable for wind turbines. If we set aside land used for urban areas, transportation, defense and industry, rural parks and wildlife, and forests, we still have about 390 million acres we could use for wind turbines. Some places will be more amenable to wind than others—because of prevailing winds and politics.

There can be no "not in my backyard" with solar and wind energy. Consider that fossil fuels are pervasive and pollute everyone's back yards—in the air, water, and soil. Over the decades, we have learned to live with a lot of changes in our landscape, from electricity lines and highways to condos and strip malls. We will also have to live with a lot more solar panels and wind turbines. The trade-off is that we'll have cleaner air, cheaper energy, and, most importantly, we will be saving that land and landscape for future generations. We will have to balance land use with energy needs.

## NUCLEAR

Nuclear energy can and does work, but 50 years of debating it have passed and we still haven't agreed on the best way to handle proliferation and waste issues. It's not "too cheap to meter," as was once predicted;[11] in fact, it is likely more expensive than renewables. The exact costs of nuclear depend on whom you ask. For instance, operating costs of a particular plant can be impressively low. On the other hand, many think the costs should include military and disposal expenses necessary to maintain a safe nuclear fleet, which significantly increases costs. There are many more examples of such conflicts, leaving the true costs a matter of considerable debate.

Nuclear has been a dependable source of baseload power, though. Baseload refers to the most reliable energy resource in a grid service area that you are least likely to lose or turn off. But experts now frequently argue whether baseload is as important as previously thought.[12] (In fact, we will discuss this in detail in chapter 8.) We likely need less baseload power than people think, and perhaps none at all, because of four factors: the inherent storage capacity of our electric vehicles, the shiftable thermal loads in our homes and buildings, commercial and industrial opportunities to load-shift and store energy, and the potential capacity of back-up biofuels and various batteries.

The approximately 60 nuclear facilities and 100 reactors in the US already provide roughly 20% (about 100 GW) of all our delivered electricity (around 450 GW.) The problem is that nuclear plants take decades to plan and build. In 2016, Watts Bar Unit 2 was connected to the grid—it took 43 years from the beginning of construction to grid connection.[13] It was the first new reactor in the US since 1996.[14] Only a relative handful of new plants are being planned, and quickly scaling up nuclear power would be difficult—due to politics, not inherent technological limits.

Another underappreciated problem is that current nuclear power plants require river or ocean water to cool them down, which ends up heating the water to levels that are deleterious to fish and plants. Two-fifths of all fresh water in the US passes through the cooling cycles of thermoelectric power plants. In many states there is not enough cool water available to install more plants of this nature. The amount of nuclear power we could potentially deploy using current technology is likely only double or triple that we already deploy, and consequently would only supply around 10–25% of our 1,500–1,800 GW target.

The United States could try to build nuclear plants faster. We could also make them cost less by changing the regulatory environment, since the interest rate on the money borrowed to build a nuclear plant can amount to significant added costs. We could develop next-generation nuclear technologies. We could use mass production techniques and economies of scale to lower their cost. But that's a lot of ifs, and it's unlikely that the country will do all of this before the combination of renewables with battery storage proves itself to be far more cost-effective and politically favorable.

Nuclear power is so fraught with problems that Japan shut down its plants. So did Germany. China is also slowing down on nuclear technology. This isn't because nuclear doesn't work (it does) but because the social, political, ecological, and economic question marks that surround nuclear make it a long, hard road to expanding the world's energy capacity. And let's not forget that it's more costly than solar.

The US Department of Energy itself has set targets of 5 cents per kilowatt-hour (¢/kWh) for rooftop solar, 4 ¢/kWh for commercial solar, and 3 ¢/kWh for utility-scale solar by 2030.[15] Still, it's unlikely we'll eliminate nuclear energy in the US, for reasons of national security. Unless we completely disarm, it's unrealistic to imagine the US pulling away from nuclear power altogether. In order to address climate change, the likeliest scenario is that we'll mildly increase nuclear (fission) power capacity in the US, but it probably won't become the dominant energy source for all the reasons I've explained. In other countries with very high population density or a lack of renewable resources, nuclear or imported renewables (most likely as electricity, or maybe as hydrogen or something similar) are the only realistic options.

All this probably leaves you wondering where I sit on nuclear. If I were king of the world, I would do without it and live more simply. Given that I can't enforce that on my fellow humans, pragmatically I think nuclear will have a place in the world's future. But I think it would be irresponsible to add more nuclear without much more investment in improving nuclear technology, waste processing, and security.

But I may be persuaded otherwise. While writing this book, I spoke with the founders of a Bill Gates–funded fusion-energy company who, like me, went to MIT. I think this company has a viable pathway to fusion energy, but they themselves admit the challenges of time and cost. If we believe their claims of 5 ¢/kWh generation and the timeline of their first installed prototype, 2032, what they're offering is still a bit expensive, and too late. I want fusion to succeed, and I think it will. But I do find that to be a slightly scary thought. My longtime friend, the wonderful thinker and author George Dyson (son of physicist Freeman Dyson), poses the question of what humans would do if energy were so cheap that we could move mountains on a whim. I worry we would dominate nature in a way that would make the world awful (think about the consequences of fusion-powered bulldozers).

## YES . . . AND

We'll need a diversity of energy sources, so stop anyone who tries to tell you about *the* answer. We can move past the arguments about how to decarbonize by embracing "yes, and. . . ." Yes, and . . . if we can make these energy technologies work at scale, we should. This applies to renewably generated liquid fuels like ammonia, airborne wind energy, low-energy nuclear reactions, cold fusion, and whatever else might emerge from similarly creative lines of thinking. Yes, and . . . if cheaper biofuels, or a synthetic fuel, or hydrogen work out as storage mechanisms, they can come to the party.

"Yes, and . . ." allows for technological advances in carbon sequestration, or fusion, or something even more incredible to emerge—if we invest in the right R&D, and if we get a little lucky. But, as I have said, it's too late and too dangerous to rely on miracles. Any precious capital going to these other projects is not going to the zero-carbon solutions that we already know work. "Yes, and . . ." avoids arguments that distract from the main players in decarbonization, while allowing that other technologies can all make small, but vital, contributions.

There is nothing that physically and technologically limits us from doing it all with renewables. There are only cynical or specious arguments that say we can't. The biggest barriers remaining have the same origin: inertia and the stubborn insistence on the current way of doing things. This manifests as fossil-fuel subsidies and massive misinformation campaigns. It's also buried in old ways of doing things, like the state-sponsored utility monopoly, which gives low interest rates to big projects instead of to consumers who need to swap their gas heaters for solar and heat pumps.

There will be trade-offs. More nuclear means fewer batteries but more public resistance and, most likely, higher costs. More solar and wind means more land use. What we cannot afford are plans that make no progress because we are wasting time arguing over these issues before we begin, or because we are over-investing in technologies that can't scale up sufficiently. The real test, given the urgency of our climate situation, should be, "Is it ready to go to scale today?"

We need to act now.

# 8

## 24/7/365

---

- Renewables are intermittent sources of energy, but they complement one another.
- Everything that can store energy should store energy.
- Every end use of energy that can be shifted to when the sun is shining or wind is blowing, should be shifted.
- By electrifying sectors that were previously not electrified, it becomes easier to balance the grid.
- We'll need to share electricity with our neighbors and borrow it back from our friends.
- We'll also need to expand long-distance transmission infrastructure to send electricity across state lines.
- Just as with fossil-fuel infrastructure, there are big cost benefits to over-building capacity.
- We critically need "grid neutrality" to allow our twenty-first-century infrastructure to offer the most benefits.

We've established how much energy we need, where it can come from, and how it will make all Americans more comfortable without giving up anything except bad air, corrupt politics, and dirty water tables. As you'll see, if we can finance it appropriately, it will also be far cheaper (chapter 10) and will provide millions of new jobs (chapter 15). **So why aren't we already electrifying everything as fast as we can?**

People who resist decarbonization often have vested interests in continuing to burn fossil fuels. Others just don't like change. These dinosaurs often wrap their opposition in a critique that renewables are intermittent, expensive, and unreliable. They say renewables are fatally incompatible with always-on, 24/7/365 electricity. Since renewables have outputs that fluctuate—based on weather patterns, the seasons, and whether it's day or night—the concern critics raise is that supply won't be able to keep up with demand, causing brownouts or blackouts.

It's true that we have come to expect that when we press a button, the stove cooks and the lights go on, and when we turn the faucet, the water is warm. Reliability was built in to the twentieth-century grid as part of a grand bargain giving a monopoly to corporate utilities in exchange for their assurances of 24/7/365 reliability and service to the under-served. This deal worked pretty well through the twentieth century, but it left us with a mixed bag of incentives that don't motivate the energy sector to decarbonize or innovate rapidly enough to address climate change.[1] Rural electricity co-ops serve another significant portion of US consumers and have their own set of challenges that likewise slow our progress toward the better world our children deserve.

Using renewable sources, we run into multiple problems when we expect always-available electricity. There's the 24/7 challenge of day and night cycles (we need light when it's dark), and there's the 365 problem of the seasons (we need more heat in the winter, just when the sun is lowest in the sky, and we need more air conditioning in the summer, just when the air around us is the hottest).

I believe that we already have the answers to these challenges; and while it will be far from easy to implement them, the solution is simpler than you think. Twenty-first-century business juggernauts were built on the commodification of logistics—and we need to do the same for our energy system. Although there is a lot of work to do, there is no reason to delay a full-speed embrace of the clean-energy future we need.

Smoothing out demand minute to minute, hour to hour, day to day, and month to month will require all the ingenuity we can muster. Fortunately, America has loads of ingenuity. We have existing ideas that will solve most of the problem. We'll also see that our connectivity to each other is critical, as averaging effects, geographical effects, and the ability to lean on one another's generation and storage capacities are the only realistic ways to ensure reliable delivery.

This book is about outlining the pathway to "yes" so we can all go fight for it, and I want to provide enough detail to quiet the plethora of naysayers. So let's look at the tricks we have to make the grid we need work 24/7/365.

This is the hardest problem remaining in decarbonization. It's not go-to-the-moon hard, it's organize-many-things-to-work-together hard. Sound familiar? That's what we did when we built the internet.

As I showed earlier, we can get the majority of the 1,500–1,800 GW of electricity to power America from renewables. It is worth reminding yourself that this means generating and delivering three to four times as much electricity as we currently do. We won't do this by tuning up the old grid; it will require rebuilding the grid with new twenty-first-century rules and internet-like technology.

## THE 24/7 PROBLEM

The contemporary American home makes use of a variety of energy services. Our needs for energy throughout the day vary. Most homes require more energy in the morning (for showers, laundry, and breakfast) than during the day. We need even more in the evening (for lights, heating and cooling, food prep, dish washing, and entertainment). Demand drops during the day when people are out—though it rises in offices and industrial buildings. While we are out of the house a few things still stay on, such as the fridge's compressor, some lights, and always-on devices like cable boxes, wireless routers, clocks, and timers. We have a big load lump in the morning, a lull in the middle of the day, and a bigger lump in the evening that reduces to a trickle overnight.

Added to hourly variation in demand, there are also daily differences caused by weather fluctuations and larger seasonal variations.

Right now, depending on where you live, these various energy demands are powered by some combination of natural gas, electricity, propane, firewood, and oil. Switching to renewables for all of these sources solves the carbon problem, but it does introduce tremendous load variability. Imagine a family that has electrified its cars, furnace, water heater, cook stove, and clothes dryer. The whole family is out for the day, and the load in the home is tiny. They arrive back in the evening, after the sun goes down. Dad cooks dinner on the electric stove at the same time that mom starts a load of laundry. One kid jumps in the shower to wash off the day's activities, and the other kid turns the thermostat up to warm up after a day outside in the snow. Both cars get plugged in for a recharge. The home that was using almost no electricity at 3 p.m. suddenly has 20–50kW of loads turned on, all demanding their share. This is the most extreme case of load variability.

The thermal loads, when electrified, are big and heavy. Culturally, we like high heat for cooking; cook-top manufacturers boast about high-BTU burners. Electric heat pumps are very efficient, but similarly have very high instantaneous electrical loads. Air conditioners, too, are notorious energy hogs. Drying clothes is also inherently a high-energy process, as it takes a lot of work to spin and evaporate all the water away. My wife and my father would think it remiss if I didn't espouse here the benefits of line drying clothes, which makes them last longer and smell better—not to mention that it's a fabulous way to utilize free solar and wind energy. Incidentally, it also highlights one of Australia's proudest inventions, the Hills Hoist outdoor clothesline (solar-powered!).[2]

We've been trained in instant refueling by our gasoline cars. If you have already moved to an electric vehicle, you already know you need a very big circuit to charge it quickly. On a typical 30-amp circuit at 120 V, you'll only get about 10 miles of range for each hour of charging. This is why people move to higher currents and higher voltages for car charging, typically 40 amps and 230 V. At these rates you can get about 25 miles of range for each hour of charging. Some people even push for 480-volt "superchargers."

I have put sensors on my current house to watch how all the loads work, and the mess that you get is shown in figure 8.1. If we look at the minute-to-minute energy use, we can see that it's completely crazy. Overnight there might be a night light on, and then occasionally my refrigerator compressor switches on, and the total load on the house is not even 100 W. However, if we are charging two cars, using the induction stove to cook dinner, running the dishwasher and clothes dryer, and heating the

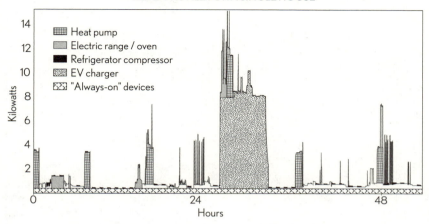

**8.1** Measurement devices such as Sense(™) can let you look in incredible detail at your energy use. This load profile, from March 10–11, 2020, highlights the challenges of our all-electric future.

house and the hot water system, the house needs something like 25 kW. Running everything at the same time is a bad idea.

Load variability is difficult for an individual house like mine. If my house were the only house on the grid, the grid wouldn't be able to meet my loads. It takes time to "spin up" a coal or natural gas plant. (Spinning up refers to literally spinning the generator that creates the electricity.) If mine were the only house, I'd flick the light switch, or turn on the TV, and wait an hour for the coal plant to be fired up and start sending me electricity. This is why it is useful to think about groups of homes, and our averaged loads. Not all homes are exactly the same, and if we start pooling everyone's homes together, where people cook dinner and shower at different times, the loads start to balance out. But while pooling evens out minute-to-minute variations, the demand still varies throughout the day (just like traffic!). There are people called grid operators who carefully plan out and manage the generation connected to the grid to give us what appears to be a constant supply of electricity. They spend their time matching supply (generation) to demand (load). When it goes wrong, we experience brownouts and blackouts. Load management is critical, both for the current grid and especially for the grid of the future. We rely

## ELECTRICITY USE OF A GROUP OF HOUSEHOLDS

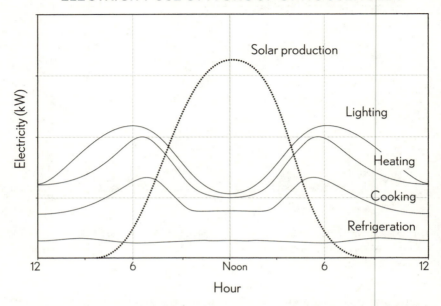

**8.2**   If we group a bunch of houses together to average out the exact moments we turn lights on and boil kettles, we get this cartoon of residential house electrical loads on which we can overlay the solar generation curve.

upon each other's loads and these averaging effects to make the grid work today, and this interconnectivity will be even more important tomorrow.

Collectively, the 24/7 variability of all our individual homes can be seen in daily demand curves, where we have a small peak in the morning, a slump in the middle of the day, a big increase in the evening hours when everyone gets home, and a drop-off through the night. This is drawn as a cartoon in figure 8.2. California has progressive energy policies, and as with many things is an early adopter of green energy. More and more people are putting solar on their rooftops and using it "behind the meter," meaning they meet the loads of their own house through the middle part of the day. This is poorly matched with the typical daily electricity consumption profile, and this mismatch manifests as the "duck" curve shown in figure 8.3, famous because it resembles . . . a duck. More and more people produce solar on their rooftops, which peaks in the early afternoon then drops off as demand rises dramatically in late afternoon

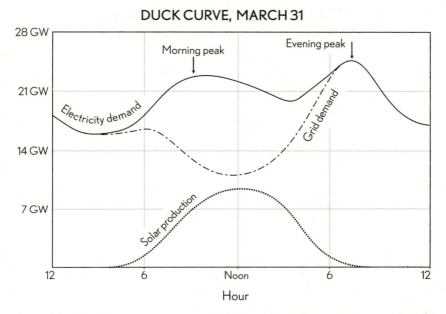

## DUCK CURVE, MARCH 31

**8.3** The "duck curve" shows the impact of behind-the-meter solar energy on grid demand in California. Each year, as more solar is installed on rooftops, the belly of the duck gets fatter.

and evening. Each year we put more solar behind the meter, the middle of the day draw from the grid decreases, and the belly of the duck gets fatter.

### THE 365 PROBLEM

The duck curve is not the only supply/demand challenge in the new world that is emerging; there is also the seasonal problem. There is more sunshine in the summer and more wind in the winter. People have known this for millennia, and it is represented in collective data we can take from all the country's wind and solar farms.

Added to that problem, we need more heat in the winter, and we like more air conditioning in the summer. It turns out that we also drive a little more in the summer, and we use more electricity for all of the other things in our house in the winter, when we are indoors more often. These

## SOLAR AND WIND VARIATION ANNUALLY

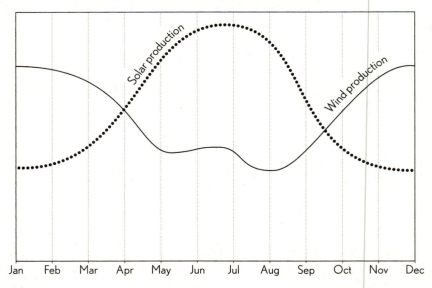

**8.4**   Solar and wind vary annually, as is intuitive to those of us who follow the seasons.

phenomena manifest themselves as the seasonal load profiles which we can determine by looking at the annual data compiled by the Energy Information Administration on electrical loads,[3] petroleum use (transportation mostly), and natural gas (heat, but also electricity).

So our 24/7/365 problem is plain to see: our loads change minute to minute, hour to hour, and day to day; and as the weather changes throughout the year, so do our seasonal loads. The question then becomes how we can deal with it.

## MATCHING SUPPLY AND DEMAND—THE SOLUTIONS

Few people, if any, really get to step back and look at the entire energy picture to address the problem of matching supply and demand. Unfortunately, it just isn't anyone's job description. Most people in the energy world look at their small piece, like transportation fuels, grid balancing, or natural-gas supply. To gain confidence that the future we seek is really possible, we need to look at all of these energy flows at once. We need to know how to convert non-electrical loads like natural gas for heating into electricity. Only then can we balance all of our energy uses. We'll see that we have existing ideas that will solve most of the problem, and we'll also see that our connectivity to each other is critical.

## BATTERIES, BATTERIES EVERYWHERE (AND NOT A DROP OF OIL TO DRINK)

Addressing the variability problem will require a lot of storage, mainly in the form of batteries. Everyone knows this, but we need to think bigger about what batteries are.

The US will have to create lots of storage for renewable energy. In our fossil-fueled world, we already have vast storage facilities, so it's something we already do at scale. Natural gas is stored in giant underground caverns. America has around 4 trillion cubic feet of natural gas storage capacity,[4] which is roughly a month's supply. The infamous Porter Ranch gas leak in Southern California, at one such storage facility, resulted in an enormous leak of methane, a potent greenhouse gas. The US strategic petroleum reserves in Louisiana and Texas contain hundreds of millions of barrels of oil—but that's only about 30 days of US consumption, a testimony to how much oil we use! Most coal plants stockpile enough coal for a month of generation.[5] These energy-storage systems are required to balance supply with demand in the face of fluctuations, whether it be a cold snap, a compromised pipeline, or an oil embargo.

The most straightforward approach to supplying reliable electricity is to build storage infrastructure that allows us to deposit extra electricity when we have it, and withdraw it when we need it.

Chemical batteries, not unlike the AAs you might immediately imagine, can store electricity directly. They are quite expensive but their costs have been falling quickly. Lithium-ion battery prices were above $1,000/kWh of storage capacity in 2010, fell to $150/kWh in 2019, and are projected to be $75/kWh by 2024.[6] As a result, large-scale deployment of batteries is becoming a realistic possibility. Chemical batteries are best at ironing out the short-term or daily variations in electricity. They're excellent at storage on the order of one hour, one day, or one week, but they won't help us store energy for winter, as they would be prohibitively expensive if we only charged and discharged them once a year—it would take 1,000 years to get the benefit of the capital investment!

Still, contemporary lithium batteries only last around 1,000 cycles. They can be pushed a little further, but even then, the cost is high, about 10–25¢/kWh for *each storage cycle*. It's important that manufacturers double or triple battery cycle life, which will make the storage cost for each cycle mere pennies per kWh.

The energy game will change forever when the combined cost of rooftop solar and battery storage can beat the cost of the current grid. My battery-bullish friends think about this as the energy singularity, the moment when batteries cost less than grid transmission. I'm a tad less bullish—it will change energy economics in a very fundamental way, but we will still need the grid and all of the other tricks in balancing it. In some markets, this moment is already here, or very near. Remember that the average US cost of grid-based electricity is 13.8¢/kWh. If rooftop solar achieves the price point it has in Australia of 6–7¢/kWh, and if batteries achieve a price point of around the same 6–7¢/kWh per storage cycle, then we will have arrived at that moment when our battery storage can beat the grid on cost, and in a way that can be built out incrementally, without massive investments. If we halve the capital cost of batteries one more time, and double the cycle life, we will be in that future. It is only a matter of time. If we move faster in this direction, we will bring the future forward and have a better climate outcome.

Given that batteries are currently expensive and will never be free, we should think about all of the other things in our everyday lives that will require batteries or can be used as batteries. The batteries in electric cars will represent an enormous storage opportunity. If all of America's 250 million vehicles were electrified, they would have the capacity of about 20 terawatt-hours (TWh), enough by themselves to balance out the daily fluctuations of our new electrified world. To follow our assumptions, go with 80 kWh for each battery, enough for around 2–300 miles of range; 250 million of them is 20 TWh. Given that we're often using our cars, we wouldn't use all of their batteries' capacity, but the grid will still benefit hugely from their contribution.

Besides America's car batteries, our 120 million homes and 5 million commercial buildings have an enormous number of hot water heaters, refrigerators, and HVAC systems, all of which can be used to store energy. This type of battery is thermal energy storage, where instead of storing electricity directly, it is converted to heat (or cold) in our refrigerators or HVAC systems. In this future, where we'll have excess (solar) energy in the middle of the day, it is critical to store that away to keep our refrigerators cold and homes warm overnight. This is not radical, nor is it expensive—people already run water heaters when electricity is cheap and store the hot water for later use.

We need to find and take advantage of as many of these opportunities as possible. For instance, an inexpensive thermal storage system the size of your clothes washer and dryer could store an additional 25 kWh per household—about another 3 TWh of electricity across the US. There are already companies that sell ice-storage systems for air conditioning. We could freeze the water when energy is cheap, and use that coldness at the hotter times of day when electricity is more expensive.

We have other types of batteries. Pumped hydro is a form of mechanical battery. These systems use electricity to pump water uphill when the wind is blowing or the sun is shining, then let it run through a turbine to generate electricity on the way back down, when the sun is down and the wind is still. Pumped hydro is cheap and can work with our existing hydroelectric infrastructure. Right now, 95% of our grid-connected battery capacity is pumped hydro. It is good for short- and mid-duration storage, but current reservoirs are not big enough to store the differences between our seasonal uses. There are other mechanisms that can store energy, such as flywheels, compressed air, and hydrogen. For a multitude of reasons, they are highly unlikely to be the major players in grid-scale storage, so we'll talk about them later, in appendix A.

Using biofuels to bridge seasonal gaps can also be significant. Take wood, the best-known biofuel. We used to measure energy in cords, a 4ft. x 4ft. x 8ft. pile of lumber. Common wisdom holds that a house needs three cords of wood for the winter. With minimal management, the average acre can sustainably produce 1 cord per year, and 1.5 cords with some effort. There'd be no winter storage problem at all if every person had 5–6 acres of forest (but there might be an air quality problem). As my dear old friend David MacKay said, "For forest-dwellers, there's wood. For everyone else, there's heat pumps."[7] I'm not proposing going back to firewood—although that can be a carbon-neutral form of winter heat if done correctly, but not for everyone, not at national scale! Thinking more broadly about winter firewood lets you imagine our substantial biological waste streams as a potential winter battery. Waste from agriculture, sewage, food, and forestry could be a "battery" that easily bridges the summer–winter divide if we were to store it as a biofuel for that occasion. These waste biofuels are a resource equivalent to about 10% of our current energy supply. To what extent biofuels become part of our seasonal battery will depend on technological, economic, and policy details.

Using all of our varieties of batteries, on the grid and behind the meter, is what is necessary to make 24/7/365 sustainably generated electricity a reality.

Storage is not the only pathway to matching supply and demand, and it alone is not enough. Two other techniques are demand-response, and over-capacity, and both will likely be cheaper than batteries.

## ELECTRIFYING EVERYTHING STABILIZES OUR LOADS

In the same way that the internet gets better with more users, balancing the grid gets easier as we electrify more things.

When the US electrifies everything, in addition to homes we are electrifying the transportation sector, the commercial sector, and the industrial sector. These sectors are even larger users of energy than our homes, and just as averaging the loads of all of our homes makes it easier to electrify everything, so too does electrifying all sectors and linking them to our new twenty-first-century grid.

When we leave our homes for the day, many of us go to our jobs in industry or commerce, and consequently we take our loads with us.

We turn the lights off in our homes but turn on the computers and cash registers and production lines at our workplaces. Taking advantage of this can further balance our loads and match them to renewables.

We need to contemplate the very important links between energy and culture and society. Because coal plants are expensive and difficult to shut down (it takes as long as 8 hours to turn them back on again), we keep them running all night. Because of this we have wound up with an excess of cheap electricity at night—something people historically took advantage of to heat their hot water. We made electrical loads after dark to consume that cheap energy. We changed our energy system and our energy system changed us. Contemplate for a second how this is partly responsible for what Las Vegas is today!

We reacted to cheap power at night by creating night shifts in heavy industry so that industry could consume that power. In a solar- and wind-powered world we will have the opportunity to rethink some of these decisions. I don't know a lot of people who love working the graveyard shift, so this could provide a benefit to many workers.

The big loads in industry and the commercial sector that can be shifted will help a lot. A huge amount of energy is used in the cold-chain, which refers to the set of refrigerated warehouses, vehicles, and other storage depots that keep our giant food supply cold and fresh. This load is shift-able without compromising any of the food; we just choose when to run the compressors that drive the refrigeration that keeps the system cold and manage the temperatures more carefully as though in an icebox. In every sector, everything that can be a battery, and everything that can shift a load, should do so.

Even our steel mills and aluminum smelters will be critical and offer shiftable large loads that can be moved to match supply. Together, the US steel, paper, chemical, and food and beverage industries consume about 6 billion kWh per day.[8] That is the equivalent of 50 kWh per household—a huge home battery. Manufacturers can still produce the same amount of goods in the long-term, but they can match their major loads to the available energy supply over time. You can imagine that maintenance might now be scheduled for the winter period when solar production is low. When there is ample energy, they can over-produce goods. It is often cheaper to store products than it is to store electricity directly. We already warehouse summer grains so that we can eat bread in the winter. We could expand this seasonality to our durable goods, offering companies cheap electricity so they can make hay when the sun shines.

## DEMAND RESPONSE: BALANCING THE LOADS

Besides storage, another tool to balance the grid is to adapt demand-side loads to account for intermittent supply. We already commonly use this kind of demand response. In our current energy landscape, electricity is cheap at night, because demand is low and it is too hard to turn off the fossil fuel plants. People set the timers on their pool pumps and hot water heaters so that they turn on at these times. In the future, the cheap electricity will be in the early afternoon because of solar—we'll just need to change the timers.

A typical house currently uses around 25 kWh of electricity every 24 hours. If you electrified the two cars in the driveway and drove each of them the American average of approximately 13,000 miles per year, then the cars' combined constant equivalent load would add an additional ~20 kWh per day. Electrifying everything currently powered by natural gas—hot water, space heating, and cooking—represents a further ~30 kWh load (provided it's done efficiently with heat pumps; otherwise it's closer to ~80 kWh). Electrifying the whole household roughly triples the amount of electricity it requires—and this will eliminate the need for gasoline and natural gas. While this might initially seem like a problem, adding thermal loads and connecting electric vehicles to the house provides greater opportunity for these machines to take turns sucking up some sunshine. This technique is called "demand response."

Many residential and commercial loads are flexible—for example, swimming pool pumps don't really care what time of day they run. By networking these devices, their demands can be timed to when the supply can accommodate them. Further, by networking across multiple houses we can ensure that everyone in a given neighborhood doesn't turn them all on at the same time. Doing this will significantly reduce the peak loads exerted on the grid, increasing reliability and offering savings in transmission and distribution.

In my own home, I am building out the system to balance all my loads. I will have the largest solar-panel system I can fit on my roof. It will produce about 20 kW nominal power (the amount of energy produced around noon on a sunny summer day). It will average about 4.5 kW through the day, and produce about 100 kWh/day. This is enough to power all my loads: an electric vehicle and several electric bikes and skateboards. The hot water heater is driven by an electric heat pump, as are all of the heating systems. The induction stove and oven are electric. All of the heating required for my home will happen in the middle of

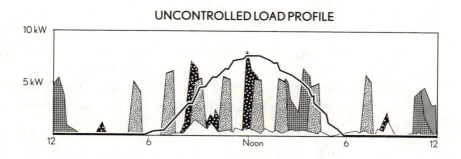

UNCONTROLLED LOAD PROFILE

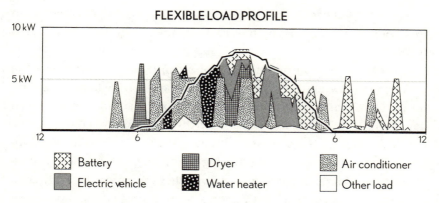

FLEXIBLE LOAD PROFILE

Battery   Dryer   Air conditioner

Electric vehicle   Water heater   Other load

**8.5**  Load profile for a "typical" house, demonstrating how demand response can move the great majority of loads into the supply curve dictated by solar energy.

the day, when solar production is the highest; heat will be stored for overnight use. The system will charge the car as the next priority, because it requires the biggest load of all. A timer can make the dishwasher and clothes washer operate during times of maximum load availability. I'll be able to squeeze the great majority of our energy into our daily solar window, but not all of it, so I'll still need a grid connection to iron out the fluctuations. I have to do some custom work on the electrical systems and controls, but these types of solutions are being developed all over the world and will only get easier and cheaper to implement. They also represent giant business opportunities to those who come up with the answers and make them simple, even invisible, for consumers to use.

In their publication "The Electrification of Buildings," the Rocky Mountain Institute explains what demand response is and what it does.

Figure 8.5 shows the before and after of demand response. Much like the load profile for my house, the original demand is a very choppy profile of electricity use throughout the day. Using as many demand-response tricks as we can, the majority of that load can be squeezed under the solar generation curve. This picture gets better the more people network their houses and the larger the pool of supplies and demands there are to share.

## YOUR WIND, OUR SUNSHINE, YOUR NUCLEAR, OUR HYDRO

The country will need lots of long-distance transmission infrastructure so that your sunrise powers my breakfast and my sunset powers your late-night TV.

If we have 10 wind turbines in California, there are days where the wind won't blow, and we don't make much power. If we have 10 in California, 10 in Idaho, 10 in Texas, and 10 in North Carolina, on any given day, there is an excellent chance that they are collectively producing power. Similarly, if it is overcast in Virginia, the sun is probably still shining in Florida and New Mexico. The bigger the geographic region we connect to the grid, the higher the likelihood that we can produce power all of the time. The contiguous 48 states span four time zones, broadening the solar window. East-Coast sunshine can help states in the middle of the country through their early morning rise in demand. Late-afternoon California sunshine can power the last demands of the evening peak in Chicago. The evening breezes over the plains can get California through the night and help the East Coast rise.

Long-distance transmission of electricity was necessary in the twentieth century because we had a hub-and-spoke model of electricity distribution. Giant generating plants at centralized locations connected to our homes via transmission and distribution lines. A new grid, with widely distributed renewables, needs this long-distance transmission even more. Keeping some of the twentieth-century generation technologies can make things easier. There are currently around 100 GW of nuclear electricity feeding the grid. This baseload resource can fill in gaps everywhere. Expanding its capacity could further ease supply anxieties around the country—but this is predicated on transmission that goes farther and carries more electricity.

Moving energy in quantity from north to south, from east to west, makes the 24/7/365 problem just that much easier. It gets easier again if Americans share with our international neighbors, with Canadian wind and Mexican sunshine helping to bolster supply. Just like the internet, the more connected we are, and the bigger the wires, the better it gets. The grid already has major interconnections that cross time zones and state boundaries. We don't need to imagine magical new technology here; we need to commit further to the things we already know how to do.

## ABUNDANCE!

Here's a radical idea for you. We have become so obsessed with efficiency and scarcity in our conversations about the future of energy that we've forgotten to imagine a world that is driven not by scarcity, but by abundance. This abundance is overcapacity, and it's something used in the current energy system; it is going to be one of the cheapest ways to provide safe, clean, reliable energy in our renewable future, too.

One example of this is natural-gas "peaker plants" in our existing energy system that generate electricity only during peak times. They spin up, for instance, in the late afternoon to meet the demand for the evening peak. They don't operate all day, so in that sense, they are underutilized. In other words, they represent an overcapacity. Another (less obvious) instance of overcapacity is the nation's automobile fleet. Suppose we could perfectly utilize all of our cars, all the time. That would mean we would need far fewer cars to meet our needs. But because our needs to move ourselves and our stuff are variable, that perfect utilization is impossible—though ride-sharing services are working toward this. Right now, if we ran all of our car and truck engines at full power at the same time, it would represent something like 40 TW of generating capacity. In reality, our cars only use about 1 TW of power, on average, so we are something like 40 times overcapacity.

So here's a crazy idea: given that wind- and solar-generated electricity are now the cheapest energy sources at about 2–4¢/kWh, instead of fretting about decreased supply during the winter, let's just design the system to meet that winter minimum, and have an oversupply and overcapacity

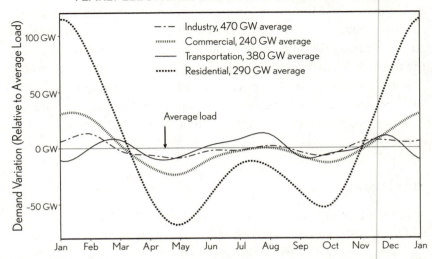

**8.6** Modeled seasonal variations by energy sector if loads were almost completely electrified.

the rest of the year. I'm not the only person thinking about this not really so radical idea.[9]

In figure 8.6, I have crudely modeled the surpluses and shortfalls over the year of an energy system where all sectors are electrified and connected, and the patterns of wind and solar generation we already see on the grid are scaled up. We see a winter peak in demand because of heating, and a smaller summer peak due to air conditioning. Similarly, we can model future electric supply by assuming current seasonal patterns of solar and wind generation. I have also used data on utility-scale solar and wind plants[10] and extrapolated the generation patterns to construct a hypothetical zero-carbon electricity supply. I include 50 times more solar capacity than we currently have and 30 times more wind. I also double the current nuclear and hydroelectric supplies. Not surprisingly, in figure 8.7, we see a summer peak due to the high degree of solar.

Conveniently, the wind blows more in winter, so these two supplies work to mostly balance each other out. Unsurprisingly, January is the month with the lowest supply, and our highest supply is in late spring when the wind is still blowing strong and the sun has reemerged.

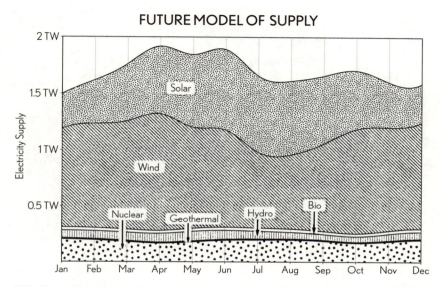

**8.7** Seasonal variations of a total US supply based on majority generation by wind and solar (using existing historical production patterns of wind and solar as the basis), with a small increase in baseload nuclear and hydroelectricity.

Putting our new supply and demand pictures together, we can examine the surpluses and shortfalls over the year. We could try to use summer excesses in supply through some magical storage technology to meet the peak winter demand. Alternatively, we could just produce much more energy than we need in the summer, and have the winter minimum of supply hit the demand maximum. We show this in figure 8.8. To reliably provide enough electricity for all demands, all year long, we'd only need to overbuild our supply capacity by about 20%. At 2–4¢/kWh for grid-scale electricity, that would only increase the cost of our generation capacity by 0.5–1¢/kWh. This is a much cheaper option than any of the batteries I have discussed above. Given that we have a pathway to 6–7¢/kWh electricity on our rooftops, and industrial wind and solar is around 4¢/kWh, it doesn't strike me as crazy that energy producers will add an extra 20% for the peace of mind it will bring. This summer energy excess will likely be absorbed in the production of hydrogen or ammonia or even in the scrubbing of carbon from the atmosphere.

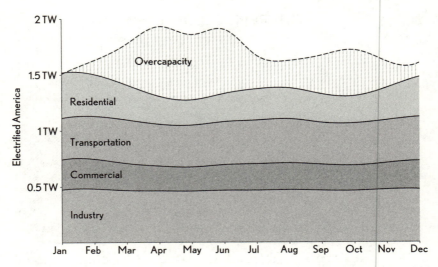

**8.8** An oversupply of energy production capacity, much as we already have today, enables us to use renewables to confidently meet our seasonal challenge of the peak winter load corresponding to the winter renewables production minimum.

We can balance our demands and our supplies, and we can do it without fossil fuels; we just have to throw out some old ways of thinking. Our clean, carbon-free future is going to be an age of abundance.

To make all of these tricks work in concert, which is enough to solve the problem, the critically missing ingredient is a grid that can tie it all together.

## A TWENTY-FIRST-CENTURY GRID

In 1973 and 1974, a small group of researchers working on ARPANET, the precursor to the modern internet, designed a set of protocols commonly called TCP/IP that determined how information would flow over the network. They invented the unit of information called a packet.

The great innovation was a protocol that ensured that all packets on the network were treated equally, no matter what data they contained, where they came from, or where they were headed. This architecture was explicitly designed to scale and to adapt to changing technology, and it did, growing from a small academic and military network into the modern internet, which comprises billions of connected devices that send and receive uncountable packets across it.

It should be our goal to enable a similarly decentralized electrical network protocol that allows the rapid movement of "packets" of electricity between billions of connected loads and uses them as needed for storage and balancing. The analogy breaks down a little because the internet can be purely digital, whereas managing the flow on the electricity grid isn't about managing discrete packets but rather voltages and currents—but is still a directionally appropriate North Star that the US should aim for. People have implemented such systems on a small scale, often calling them micro-grids, but fully electrifying the US's energy system will require the creation of a decentralized network of all the energy supplies and loads in a plethora of overlapping, connected micro-grids.

We could get to a point at which we can truly share, at scale, all of the demand-response possibilities, and all of the storage and battery opportunities in all of our homes and vehicles. Small amounts of storage everywhere add up to the giant battery we need.

Right now, if you have solar panels, you may sell some of your energy back to the grid, but often with caveats, such as the requirement that you can only sell back as much as you use so the accounts balance to zero at the end of the month. We need to make it universally possible for households to connect as much solar and storage as they like. Similarly, the policymakers need to empower citizens to offer their vehicles and appliances as part of a connected, nationwide demand-response infrastructure. Our future system must be more innovative than time-of-use pricing and more flexible than net-metering. We need a grid that treats everyone connected to it as both a supply and a demand, a load shifter and a battery.

Let's start demanding **grid neutrality**. Join the board of your rural co-op. Write your representatives. Get elected to a state utility commission.

# 9

# REDEFINING INFRASTRUCTURE

---

> ☞ Americans need to stop imagining that focusing on small "green" consumer purchases will save the planet.
> ☞ Instead, we need to focus on a small number of big purchases that define our personal infrastructure.
> ☞ Our personal infrastructure will be critical to the shared infrastructure of our twenty-first-century energy system.
> ☞ Redefining infrastructure allows us to think about new ways of financing these purchases.

Infrastructure comprises the basic physical and organizational structures and facilities required for the operation of a society or enterprise. We currently think of infrastructure as mainly dams, roads, rails, and bridges. But to build a clean economy, we need to expand this definition of infrastructure for the twenty-first century.

Where twentieth-century infrastructure largely emphasized a supply-side view of the world, the twenty-first-century infrastructure encompasses the demand side as well. It is not just the roads that matter, it's the vehicles on them and the batteries inside those vehicles. It is not just where the transmission lines go, but what is wired up at the end of them: the water heaters, ovens and stoves, heat pumps, and refrigerators. And the end consumer is not only connected to the grid, but also to everyone around them.

We need to redefine infrastructure for three reasons. The first is that it helps us as individuals focus on the big things that will move the needle substantially with respect to our personal $CO_2$ emissions. The second is that it enables us to clearly see the connection of our personal things (furnaces, cars) to our collective things (the grid, transmission lines). Thirdly, and most importantly, it allows us to think about new ways of financing these purchases.

## YOUR PERSONAL INFRASTRUCTURE

Redefining infrastructure starts with our personal infrastructure, which is what we need to pay attention to in order to help address climate change. What makes up our personal clean-energy infrastructure are the machines and appliances we use daily—though they are often invisible to us—that determine our large carbon footprints. These big-ticket items in our households—cars, furnaces, water heaters, stoves, and dryers—along with our decisions about what to fuel them with, drive more than 40% of US emissions today. If we add small business and commercial business decisions around these same things—how we heat our offices and what fuels our company cars use—we are making the choices that create more than 60% of our emissions. This is why we need to understand each of these purchasing decisions as infrastructure. We can make them well and have a huge impact on our emissions.

Environmentally concerned citizens today pay a lot of attention to small daily purchasing decisions and make complicated moral calculations about grocery bags, synthetic meat, vacation flights, and plastic packaging. Of course, every little bit counts, but this thinking, as we've seen, is mired in the efficiency framework of the 1970s (Reduce! Re-use! Recycle!). These purchasing decisions make a little difference but aren't ambitious enough to solve our larger carbon problem.

We need to think bigger. We need to fundamentally rethink our infrastructure so that doing the right thing is baked into the world we inhabit. If America designs its infrastructure right, we will be able to live our lives without sweating all the small things.

We need to start prioritizing the big, infrequent decisions that really matter to the decarbonized future. Where do we live? How do we get around? What do we drive or ride? How big is our house? What's on our roof? What's in our basement? What appliances are in our kitchen? Are they all electrified?

If you invest in the right personal infrastructure, you can be part of the solution to climate change merely by waking up and going about your daily life.

To be a good climate citizen, you just need to make four or five big decisions well. These are purchases (or investments) that are made roughly every decade—what's in your garage, on your rooftop, and heating your house. Choose wisely, and you can pretty much forget about the day-to-day hand-wringing. These infrequent decisions are the ones that lock us into using either a lot of energy or a little, and spewing carbon dioxide or not.

Here are the main purchasing decisions climate-conscious people should worry about:

1. **Your personal transportation infrastructure**: Everyone's next car, and every subsequent car, should be electric. (Of course, public transit, bicycles, electric bicycles, electric scooters, or anything that isn't powered by fossil fuels are even better options.)
2. **Your personal electrical infrastructure**: Everyone should install solar on their roofs at the next opportunity, whether that be a retrofit, replacing shingles, or when buying or building a new house. You should be installing enough solar to power your electric vehicles and electrified heating systems, not just the small solar systems of today that only accommodate your existing electrical load.
3. **Your personal comfort infrastructure (HVAC)**: Replace furnaces and gas- or oil-fired heating systems with electric heat pumps. Additionally, it is wise to insulate and seal homes. If you are replacing your flooring, it is a perfect time to install radiant hydronic heating systems. Choose efficient air conditioning, and buy systems that allow you to heat and cool only the rooms you are in, instead of the whole building.
4. **Infrastructure in your kitchen, laundry and basement**: Choose the most efficient and electric refrigerators, dryers, stove-tops, ranges, water heaters, and dish and clothes washers that are available.

5. **Your personal storage infrastructure**: As the country becomes increasingly electrified, there will come a moment when a small home battery will make economic sense to install as a backstop to personal energy demands (and this will also make the grid more robust). We don't need to argue; in the spirit of "yes, and," there will also be grid-connected batteries. The point is that there is enough storage required that everyone needs to participate. Cost will be the ultimate decider, and I'm going to bet we'll do more storage closer to the end use, because then transmission and distribution costs will be cheaper.

6. **Your community infrastructure**: Support clean-energy infrastructure in your community and state, so that all of your personal infrastructure is connected to carbon-free electricity sources. Advocate for solar cells over your school and church's parking lots.

7. **Your personal dietary infrastructure**: It is not as obvious to think about your dietary choices when discussing infrastructure, but the decision to eat less meat, or become vegetarian or even vegan, is one with a very high impact on your energy and climate emissions. While strict vegetarianism is not necessary, a decision to shape your diet in line with a hot, crowded planet has positive impacts for you and the environment.

If we all make these choices, this will go a long way toward addressing climate change in our own lives and in our communities. We also need to lobby our landlords, friends, and family members to make these same choices. Think about the potential of our personal infrastructure at scale.

## CONNECTING PERSONAL INFRASTRUCTURE TO OUR COLLECTIVE INFRASTRUCTURE

Understanding the connection of our personal infrastructure to existing infrastructure allows us to see our homes and cars as the batteries that are critical to this neighborly infrastructure in a clean, electrified world. Of course, the US can't get to a zero-carbon world purely through citizens' personal consumer decisions—we critically need government and industry. But the easiest emissions for us to eliminate as individuals are those we directly control as everyday consumers.

The last few decades have seen the rise of the sharing economy; people can rent out their homes or rooms with Airbnb, and shared bicycles and scooters are now part of our transportation infrastructure. Everyone's internet is richer and better because we all contribute content through Instagram and YouTube.

Americans need to get comfortable with the fact that balancing our whole energy system is going to rely on shared infrastructure, which emphasizes the need to write the rules of the road for how all of those things connect carefully and without bias. We could of course do it all individually, with everyone trying to buy a big enough battery to handle their own loads, but that would be the most expensive way to decarbonize. Clever and connected personal and community infrastructure is the key to reducing costs for everyone.

## THE FINANCING JEDI MIND TRICK OF NEW INFRASTRUCTURE

How we are going to pay to electrify everything is going to be hugely important. The personal infrastructure of our lives is critical to the infrastructure of the twenty-first century, so it should be accessible to everyone at the lowest costs possible, including financing. If a small portion of my car battery will be used to balance the grid, and occasionally the grid uses my heat pump and hot water heater to shift loads, then why should I pay retail credit-card interest rates on those objects instead of low interest rates more appropriate to infrastructure?

Redefining infrastructure allows us to contemplate the intriguing notion that the US might be just an interest rate away from a climate cure. As we'll see in the next chapter, lowest-cost infrastructure-grade financing is crucial. These pieces of personal infrastructure are individually expensive, and very few people will be able purchase them in cash, so financing will be the key to cost effectiveness.

We need to turn the climate conversation to fixing both our industrial infrastructure *and* the infrastructure of our lives. If we build poor infrastructure or make poor choices at critical purchasing moments, then we will lock in undesirable carbon output, and we will fail. If America builds good infrastructure and supports good decision making, we'll all be able to live well, use energy effectively and efficiently, and address climate change without having to think about it every minute of our lives. I am reminded of my late friend David J. C. MacKay's maxim "every big thing counts" in his wonderful treatise on energy, *Sustainable Energy without the Hot Air*.

It's the 2020s, and with a twenty-first-century definition of infrastructure, we can see our way to a clean, electrified world.

# 10

## TOO CHEAP TO METER

---

> ☞ Technological improvements in the last two decades have reduced the cost of critical technologies—solar, wind, and batteries—to below that of fossil fuels.
>
> ☞ The scale of a project to decarbonize the US is sufficient to drop the cost of renewables by half, such that they trounce the cost of fossil fuels.
>
> ☞ We need to look at the total cost of electricity, including transmission and distribution, not just the cost of generation.
>
> ☞ The cheapest energy system will maximize household, local, and community generation and blend it with industrial renewables.

We have the technology to create a carbon-free future, but can we afford to make the switch? It seems sacrilegious to discuss costs when considering the future of our planet, our species, and the beautiful critters and plants we share the earth with. It's dismal to have to justify the "economic cost" of doing the things that will make our future better. But I will sharpen my pencil and show you how, in fact, the carbon-free future will save everyone money.

We have the opportunity to solve climate change *and* make energy cheaper in the future.

## ELECTRICITY IS CHEAP, AND GETTING CHEAPER

Already, generating clean electricity is extremely cheap, and getting cheaper, and some of it will be cheaper still when it's behind the meter—provided policymakers don't screw up by implementing the wrong rules and regulations, as I will discuss in chapter 14.

When energy nerds compare the prices of different types of energy, they talk about the levelized cost of energy (LCOE). This is how much a particular technology costs per kWh of produced electricity when all lifetime costs are taken into account (such as building, operating, and decommissioning a plant). The same energy nerds talk about the capital costs of energy equipment as $/W. The asset-management firm Lazard, which tracks LCOE to guide investments, has data showing how much cheaper renewable energy sources are compared to fossil fuels.[1] The latest report places utility-scale solar at ~3.7¢/kWh and wind power at ~4.1¢/kWh. Compare this with natural gas, which clocks in at ~5.6¢/kWh and coal at ~10.9¢/kWh.

These impressively low LCOE numbers apply to utility-scale installations. Oddly, though, rooftop solar can be even cheaper because if you're generating electricity yourself you don't have to pay for distribution. We haven't yet realized this potential in the US, but Australia has lowered the cost of rooftop generation so much that their "behind-the-meter" energy—the energy they generate on their own rooftops, without relying on a utility—is cheaper than the cost of distribution alone from a centralized plant. The average cost of distribution in the US is about 7.8¢/kWh—higher than the 6–7¢/kWh which is LCOE of rooftop solar in Australia. The Australian government subsidizes the already low cost of $1.20/W for installation by another 30–50¢/W, which means it is now installing at prices of 70–80¢/W. This gives an LCOE below 5¢/kWh! America can't make all of the energy we'll need in the future this way, but we can make an awful lot of it.

A friend and fellow Aussie expat, Andrew "Birchy" Birch, wrote an influential article about replicating the Australian model of rooftop solar in the US. He showed how the dominant portion of the rooftop-solar costs in the US are "soft costs," or those not directly tied to a piece of hardware. These include permitting, inspection, overhead, transaction costs, and sales.

The Department of Energy agrees with him, and the aim of their $1/W solar moonshot is to eliminate soft costs.[2]

In the US, a solar installation happens like a custom home construction project, requiring several layers of design, specification, and oversight for each piece. Each step of the project must be evaluated and approved, incurring costs, and over the course of the process, these really add up. Taxes, overhead, and other indirect costs mean that consumers in the US are paying close to or above $3.00/W. I have colleagues—Todd Georgopapadakos, Mark Duda, Eric Wilhelm—who are working on a set of relatively simple technologies that can make this process more like installing a consumer appliance such as a water heater or electric dryer. If the US can automate many of the inspection and approval steps currently required, this will drastically lower the cost. This is just one among a grab-bag of regulatory problems that bedevil people working in clean tech in the US. These frictions prevent everyday Americans from accessing lower-cost energy.

In Australia, rooftop solar installs at under $1.20/W. In Mexico it is around $1.00/W, and in Southeast Asia it is less than $1.00/W. This is proof that the right building codes, training programs, and regulations can bring the soft costs down. There are also differences due to relative labor costs in each country—though Australian solar installers get paid around $40 per hour, which is more than double a minimum-wage job in the US.

Here is the transformative point about rooftop solar: because there are no transmission and distribution costs, it can be phenomenally cheap. Even if the cost of utility-scale generation were free, we don't know how to transmit it to you and sell it to you for less than the cost of rooftop solar. This doesn't mean the whole world will run on solar and distributed resources, but it does mean that if we are looking to make the lowest-cost energy system, an awful lot of America's energy will come from our rooftops and our communities.

## RENEWABLES ARE GOING TO GET EVEN CHEAPER AGAIN THANKS TO TECHNOLOGY INNOVATIONS AND PRODUCTION SCALING

Wind and solar are getting cheap so quickly that it's even hard for innovators to keep up. In 2006, I started a kite-powered wind-energy company called Makani Power. The idea was to produce wind energy at 3–4¢/kWh, cheaper than natural gas and 5–6 times cheaper than other wind-powered electricity at the time. The project was truly awesome. We built wings the size of 747s and tethered them with a giant cable; it flew in circles at 200 mph, undergoing 8 Gs of acceleration while producing megawatts of electricity. With investments from Google, Makani Power followed an exciting development trajectory to make our technology a reality, culminating in an offshore deployment and a demonstration in Norway in 2019, in partnership with Shell.

In the meantime, however, the wind industry at large has also made historic strides, and is now routinely deploying turbines at 4–5¢/kWh. In 2020, Makani shut down due to this evaporated advantage. Makani's technology and execution were sound, but the industry found its own way to slash costs simply by deploying at massive scale. Despite the fact that Makani's technology didn't win the cost battle, it was part of an enormous movement and ecosystem of global innovators responsible for driving down costs and making wind, solar, and batteries competitive with fossil fuels.

In 2011 I started another company, Sunfolding, with Leila Madrone and Jim McBride. We initially built tracking devices—machines that make sure the solar follows the path of the sun through the sky accurately. Our goal was to focus on solar thermal—using lots of reflected and concentrated sunshine to heat a molten salt, which heated water and created steam to generate electricity. But the relentless march of photovoltaics' (PV) price improvements beat us out of that game, and we "pivoted" (as they say annoyingly in Silicon Valley) to tracking devices for PV. We are still in the game, and we now sell our technology to industrial solar plants at basement-level prices that come out at around 2¢/kWh—lower than we ever imagined, and far lower than any fossil-generated electricity.

There are two ways to reduce the cost of energy. One is inventing the better mousetraps; the other is producing mousetraps in gob-smacking quantities. The first, called "learning by researching," is typically measured by cumulative R&D investment. The second component is caused by scaling, or "learning by doing," and is measured by cumulative total production. Makani was about an entirely better mousetrap, but it couldn't make mousetraps in quantity. Sunfolding represented one of

Price of Model T. 1909–1923 (Average List Price in 1958 Dollars)

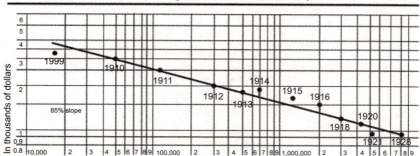

**10.1** Learning curve of Ford's Model T. Source: *Limits of the Learning Curve* by William Abernathy and Kenneth Wayne, Harvard Business Review, 1974.

many small component improvements. It was an invention, but it wasn't the whole mousetrap; it was like a better mousetrap spring. Sunfolding's tracking technology is good for taking 5–10¢/W out of the roughly $1/W. Half of this cost saving came from the hardware we invented, but, crucially, the other half was a result of reduced installation labor costs. It is these small efficiencies in materials and labor that typify learning-by-doing cost savings. As these anecdotes illustrate, and as empirical studies have shown,[3] we must invest heavily invest in *both* of these capacities to maximize long-term cost reductions in zero-carbon energy.

It is learning by doing that gets the jobs done most predictably. As we've seen, the solar and wind industries are improving, getting cheaper and cheaper with every generation of innovation. Learning-by-doing improvements are characterized by "learning rates," defined as the percentage the price falls after the production of a technology has doubled.

Among the first observations of these learning rates were Wright's Law governing the cost of airplanes.[4] We can apply this to automobiles by tracking the decrease in the price of Ford's Model T as production increased, as shown in figure 10.1. Moore's law,[5] the jaw-dropping exponential increase in integrated circuit density, can also be viewed as a version of this same idea.[6]

In the case of electricity generation, solar PV is learning at a rate of about 23% and wind at about 12%[7]—as fast or faster than fossil fuels during their early twentieth-century cost-reduction heyday. For solar,

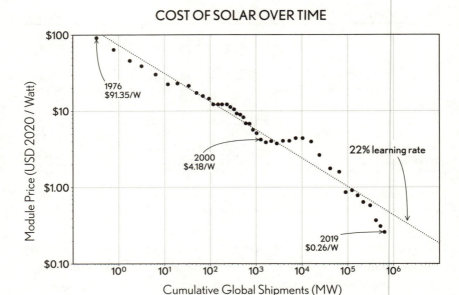

**10.2** Learning curve of photovoltaic module price. Source: Nancy M Haegel et al., "Terawatt-Scale Photovoltaics: Trajectories and Challenges," *Science* 356, no. 6,334 (April 14, 2017): 141–143, https://science.sciencemag.org/content/356/6334/141.summary.

this approximate 20% reduction in module cost per doubling of installed capacity has become known as Swanson's Law, after Richard Swanson, the founder of SunPower Corporation.[8] The progress made by this learning is plotted in figure 10.2,[9] showing how, despite extreme economic events (like the 2008 recession), solar PV modules have continued their march toward lower costs. Not only that, but in just the past five years, across the world, new installations of renewable energy have outnumbered the installations of fossil-powered energy (growing to nearly two-to-one in 2018).[10] This shift means more opportunities for learning and for falling costs.

Currently, about 250 GW of wind and 125 GW of solar are installed around the world. To reach the fully electrified version of the world, we will need about 10–20 TW of electrical power.[11] That means the cumulative production of solar panels and wind turbines still require 4–5 doublings in scale each in order to reach the annual production capacity we need. Given these known learning rates and the scale of growth required, there is ample opportunity to bring costs down even further than they've already fallen, making them even cheaper still than their fossil competition.

Pause on that thought for a moment. If we commit to wind and solar at sufficient scale to address climate change, that commitment alone will likely halve the cost of renewables again—a nail in the coffin of fossil fuels. Electricity will finally (well, almost) be "too cheap to meter," as they used to say about nuclear power.

All of this represents a rare opportunity for industry, small and large. The myths of Silicon Valley hold that disruption is always good and that progress is made by unconventional founders turning the world on its head. That model has worked in software, but in hardware, especially in infrastructure, it doesn't really work. These fields are naturally conservative, due to the graver consequences of failure and the need to guarantee machines that work reliably for over 20 years. As we've seen, progress is predictably achieved through consistent investments in research, coupled with manufacturing at massive scale. We need start-ups to innovate—and we even need crazy, breakthrough ideas, if only to inspire us to think bigger—but what we critically need is large companies to seize on these innovations and scale them up. An ambitious mobilization plan can exploit industrial learning rates to continue to bring down costs and improve the economics of the electrified future. The question is, does the US have the industrial muscle memory—or the will—to make the electrified future a reality?

# 11

## BRINGING IT ALL HOME

---

> ☞ When energy is cheap, everything gets cheaper.
> ☞ Renewables are cheap, but they have higher up-front costs than fossil-fuel technologies.
> ☞ Transitioning to renewables today will require an investment of about $70,000 per American household.
> ☞ The right policies and market scale can reduce this to under $20,000 by 2025.
> ☞ When the US decarbonizes, it will save every household thousands of dollars per year in energy costs.
> ☞ In order to finance cheaper energy for all American households, we need to create new kinds of financing.

When electricity is cheap, that makes a lot of other things in your life and home cheaper, too. Proposals to fight climate change (like the Green New Deal) are wonderful in their audacity, but they all seem to suggest that decarbonizing our lives will cost billions or trillions of dollars.

What if, instead, we start by thinking about what is required to make the clean-energy future cost less, save people money, and be an easier "sell" to the general public—the skeptics and the believers, the rich and the poor?

My colleague Sam Calisch and I built a model of decarbonization from the kitchen table outward, accounting for all of the uses of energy in our households.[1] This model shows all the money the country could save in the process of fixing climate change.

How much will transitioning to clean energy cost your household? What I'll do in this chapter is show you that it doesn't have to cost you; it can save households a big chunk of change per year, but we can only arrive at that point if we can pull all of the levers available to us. This is more than a technology problem, more than a policy or political problem, and more than a finance problem. It is a problem of all three at once.

Here is the model Calisch and I have built:

1. We use recent patterns of energy use in households, and recent energy costs, to establish the current energy costs and hence the monetary costs of running our households.
2. We determine exchange rates (of sorts) that enable us to translate our current household fossil-fueled activities into virtuous, decarbonized electrical activities. We don't change your lifestyle, we electrify it.
3. We build a simple model of what electricity could cost us in the future, given what I have explained so far in this book.
4. This is enough to calculate how much all of our household activities will cost in the future, compared to today.
5. We need to spend money to get to that bright, shiny future by purchasing the necessary CAPEX (capital expenditures or machines). These are your solar panels, electric vehicles, heat pumps, batteries, and more. From this, we build a model of the costs of your new household infrastructure and add them up.
6. The final challenge is building a financing model for our clean machines and finding whether there is an interest rate at which a household's annual payments in the electrified future are lower than our annual payments in fuel costs if we continued living as we do today.
7. Not to ruin the punchline, but the good news is, this would save us all money.

## A BASELINE OF CURRENT HOUSEHOLD ENERGY COSTS

We must first start with an estimate of current household consumer expenditures on energy. In figure 11.1, we can see that in 2018 the post-tax expenditures per household were $61,224, of which $4,136 (close to

7%) was spent on energy. The $1,496 we spend on electricity is more than we spend on education ($1,407); the $410 we spend on natural gas is more than we spend on dentistry ($315); and the $2,109 we spend on gasoline and diesel is more than we spend on fresh meat, fruit, and vegetables ($1,817).

While similar, all households aren't the same, as you can see by looking at the state-level expenditures that the Bureau of Labor Statistics (BLS) collates for California, Florida, New Jersey, New York, and Texas.[2] Households are broken down into quintiles by income. There is significant cost difference by household as a function of what they earn. Proportionally, low-income households spend roughly twice as much as high-income households on energy (6–10% for low-income households, and 5–6% for high-income households).

Given that America is so diverse, we have done the analysis for households in every state. This illuminates and lends color to the analysis, as we can see the differences in cold places and warm places, in cities where people don't drive a lot and in rural areas where people do.

We estimate all of the fuel costs by household. This includes gasoline for transportation (for simplicity, we include both diesel and gasoline under this heading); natural gas, propane, and fuel oil for heating systems; and electricity for lights, appliances, and everything else.

The State Energy Data System (SEDS) keeps detailed energy data by sector and by state.[3] This conveniently includes all residential fuels and electricity, but, critically, it does not include gasoline consumption by household. For this, we turn to the National Household Transportation Survey (NHTS). We do this state by state, and tally up the average household's total energy use. In figure 11.2, we have a baseline for our current costs, which we will use for comparison when we look at the cost of our household electrification.

## ENERGY EXCHANGE RATES

Electricity is the great energy equalizer, and the most versatile of all of the "fuels" we use. This is underappreciated. It is a bad idea to use gasoline to make light, it is virtually impossible to run an air conditioner on natural gas, and it takes a number of modifications to run a vehicle on propane.

# US AVERAGE HOUSEHOLD SPENDING

| | | | |
|---|---|---|---|
| **Personal taxes, $11,394** | State and local income taxes, $2,284 | | |
| | **Federal income taxes, $9,031** | | |
| **Savings, $3,368** | Change in securities, $1,918 | | |
| | Change in value of savings, checking, money market, and CDs, $1,449 | | |
| **Average annual expenditures, $61,224** | **Personal insurance and pensions, $7,295** | **Pensions and Social Security, $6,830** | Deductions for Social Security, $5,023 |
| | Cash contributions, $1,887 | Cash contributions to church, religious organizations, $789 | |
| | Miscellaneous, $992 | | |
| | Education, $1,407 | College tuition, $798 | |
| | Personal care, $768 | | |
| | **Entertainment, $3,225** | Pets, toys and hobbies $816 | |
| | | AV equipment and services, $1,029 | Pets, $662 |
| | **Healthcare, $4,968** | Medical services, $908 | |
| | | **Health insurance, $3,404** | Medicare payments, $565 |
| | | | Commercial health insurance, $662 |
| | **Transportation, $9,761** | Other vehicle expenses, $2,859 | Vehicle insurance, $976 |
| | | | Maintenance and repairs, $889 |
| | | Gasoline, other fuels, oil, $2,108 | **Gasoline, $1,929** |
| | | **Vehicle purchases (net outlay), $3,974** | Cars and trucks, used, $2,083 |
| | | | Cars and trucks, new, $1,825 |
| | Apparel and services, $1,866 | Women and girls $754 | |
| | **Housing, $20,090** | Household furnishings, $2,024 | Fuel oil and other fuels, $129 |
| | | | Natural gas, $409 |
| | | Household operations, $1,522 | Water and other public services, $613 |
| | | **Utilities, fuels, and public services, $4,048** | Telephone services, $1,407 |
| | | | Electricity, $1,496 |
| | | **Shelter, $11,747** | **Rented dwellings, $4,248** |
| | | | **Owned dwellings, $6,677** |
| | Alcoholic beverages, $582 | | |
| | **Food, $7,923** | Food away from home, $3,458 | Meals at restaurants, carry outs and other, $2,957 |
| | | **Food at home, $4,464** | Fruits and vegetables, $857 |
| | | | Meats, poultry, fish, and eggs, $960 |

**11.1**  2018 BLS Consumer Expenditure Survey breakdown of household spending.

Yet electricity can run all of these machines and more. It is the lingua franca of energy forms. In combination with its efficiency, this flexibility is one of its huge advantages as we look forward to a decarbonized future.

But to understand cost comparisons we need to convert transportation fuel costs into electric vehicle costs, heating fuel costs into electric heating costs, and the leftover household fuel costs into all electric ones.

## MILES PER GALLON (MPG) TO MILES PER KILOWATT-HOUR (MPKWH)

It is tricky to convert between MPG and MPkWh on the basis of energy content of fuels, because you also need to know a lot about the efficiency of each vehicle and all of its components. Fortunately, there are enough electric vehicles on the road now, and certainly enough internal combustion engine vehicles on the road, that we can use real-world mileage to convert gallons of gasoline to kWh of electricity. Calisch and I did the calculations for the same number of miles traveled in vehicles of similar size and performance capability.

For a rough sense, a small, efficient electric vehicle, like a Tesla model 3 or BMW i3, uses about 250 Wh/mile tootling around a city. That's ~4 MPkWh. The equivalent internal combustion engine (ICE) vehicle, like a Honda Civic gets an Environmental Protection Agency rating of 36 MPG average.[4]

A larger, heavier, faster electric vehicle, like a Tesla model S, uses closer to 333 Wh/mile, or ~3 MPkWh. That would compare to a larger luxury car, like a BMW 5 Series, that gets about 26 MPG.[5]

Pickup trucks and SUVs comprise nearly half of America's auto fleet. An electric equivalent, like a Rivian truck, will need around 500 Wh/mile. That's ~2 MPkWh and will compare to similar-sized trucks that get around 15–20 MPG.[6]

Using the Small, Medium, and Large vehicle models defined above, we can now translate most vehicles between MPG and MPkWh, which will give us a multiplier (kWh:G) that converts household gallons of gasoline to required kWh of electricity. As shown in table 11.1, this number is surprisingly similar (in the range of 8–9) for each of the vehicle sizes we consider. This is convenient, as it allows us to use the average value of 8.5

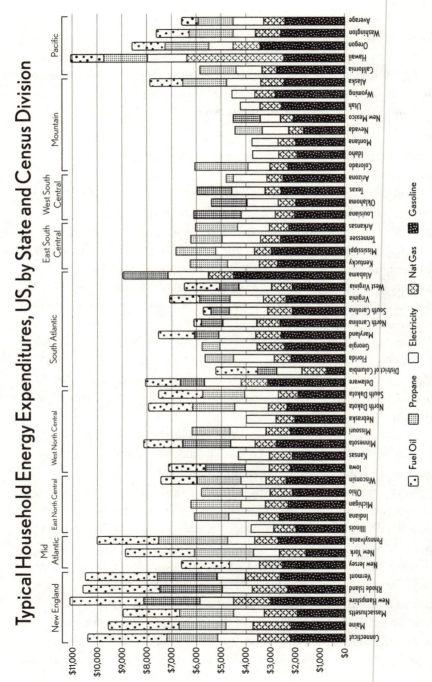

11.2  Household expenditures on energy, by fuel, by state and census division.

**Table 11.1**   ICE-to-EV Equivalencies

| Vehicle Size | MPG | MPkWh | ICE vehicle | EV | kWh:G |
|---|---|---|---|---|---|
| Small | 36 | 4 | Honda Civic | Tesla Model 3 | 9 |
| Medium | 24 | 3 | BMW 5 Series | Tesla Model S | 8 |
| Large | 17 | 2 | Chevy pickup | Rivian | 8.5 |
| Average | - | - | - | - | 8.5 |

to convert all of the household gallons of gasoline consumption to the electric equivalent, regardless of what's parked in the driveway.

## CONVERTING THERMS OR BTUS OF HEAT TO KWH

Computing the energy used for heating is more complicated than energy used for vehicles, for two reasons. The first is that not all homes are currently heated in the same way. Most homes are heated with natural gas, but many use electric and some use propane or fuel oil. The second complication is that our model largely takes account of replacing various pieces of heating equipment with electric heat pumps. The Coefficient Of Performance (COP) of the heat pump is determined by the type of heat pump (air-sourced or ground-sourced), as well as the local ground and air temperatures. We make the simplified assumption that air-sourced heat pumps are used for all retrofits, as they are lower than ground-sourced equivalents in capital cost and retrofit cost. (Ground-sourced heat pumps can have higher COP in certain regions that need a lot of heating, like New Hampshire, and may be the best economic choice in those regions.)

The annual climate model we use for each state is based on data from the approximately 1,000 TMY3 (Typical Meteorological Year) weather stations that the National Renewable Energy Laboratory (NREL) has around the country.[7] The temperatures from that data can be combined with the technical performance data of a typical air-sourced heat pump[8] and the residential hourly load profiles for all TMY3 locations as calculated by the Office of Energy Efficiency and Renewable Energy (EERE),[9] to produce an annual average heat pump COP by state for both space and water heating.

Fortunately, the EIA, in partnership with the Census Bureau, also keeps excellent data on home heating equipment type by census regions and divisions. They also keep track of the proportions of each—the percentage of homes with natural gas or fuel oil, for example.

We take existing home heating patterns for electricity, natural gas, propane, and fuel oil and convert them to kWh by dividing the current use (in kWh equivalents) by the increase in COP gained by replacing existing heating technology with heat pumps.

Our electric heating exchange rate can't be expressed as a simple ratio applicable across the US, as it was for cars, but fortunately spreadsheets and databases can take care of the accounting for all of the states and COPs. If you want to do these calculations in your head, the ratio is approximately 3.

## CONVERTING FUELS THAT AREN'T USED FOR SPACE OR WATER HEATING TO ELECTRICITY

Aside from heating, your house uses small amounts of carbon-based fuels for other activities, mostly cooking. In our calculation, we convert these remaining fuel uses to electricity with a COP of 1. Stovetops and grills are things you can't use heat pumps for, but you can use electric induction or resistance heating.

There are yet further energy costs we have to account for in the form of existing non-heat electrical loads. These are your lights, TVs, cellphones, computers, fans, pool pumps, and power tools. We assumed no efficiency wins for those loads and that they will be similar in our decarbonized world to what they are today.

## CAPITAL EXPENDITURES—UPGRADING YOUR INFRASTRUCTURE!

Obviously, we can't just plug in our current ICE cars and furnaces to a 110 V outlet and realize all of these savings. We need to buy new infrastructure for our lives—new cars, furnaces, and water heaters. What does that look like?

There are eight categories of things in the majority of our homes (we are more alike than we are different!) that we will need to fully decarbonize. We list these in table 11.2.

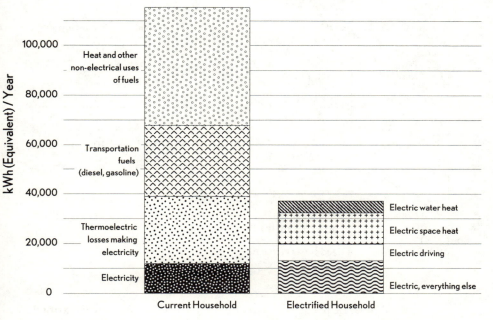

**11.3** Comparing all current energy uses in kWh equivalents to current electrical load, and the total electrical load if we electrify all current energy uses of the household.

I don't choose the luxury version of anything, but rather the average version. I also only count the difference in cost between the new thing and the old thing it replaces, as I assume you would already have cooktops and cars and space heaters anyway, so it's only fair to finance the portion that is different. Mid-priced electric ranges, for example, are about $500 more expensive than their natural gas-fired counterparts.

We add a new load center—that is the big box of wires and breakers that connects your home to the utility's meter—because your home will roughly double its electrical load, so your load center will need an upgrade. We add an EV charger for each vehicle in the house according to average state vehicle ownership per household (2.1 on average). We provide for around four hours of home battery storage to help smooth out the loads. We also electrify the space heater, and scale the capital cost proportional

**Table 11.2** There Are 9 Capital Items Considered in the CAPEX/Infrastructure Model

| Item | Typical Value | | |
|---|---|---|---|
| CAPITAL COSTS | Basis | Typical value | Financing Per |
| Range | 1 each home | $500 | 15 |
| Load center | 1 each home | $500 | 20 |
| EV chargers | 1 per vehicle | $500 (ea) | 15 |
| EV batteries | Per kWh of vehicle(s) battery | $100/kWh | 7 |
| Home battery | Hours of storage required, new electric load | $100/kWh | 10 |
| Space heater | Sized proportional to current space heating load | $5,000 | 20 |
| Water heater | Sized proportional to current water heating load | $600 | 15 |
| Rooftop solar | Sized to service a percentage of your load | $15,000 | 25 |

to the amount of heat you use. We also scale the cost of a new electrified water heater proportional to your current water heating load.

The two most expensive components we save until last. We will finance just the batteries in the electric vehicles in the driveway, as that is the principle difference in cost to an ICE. We size the batteries according to a national average vehicle range (250 miles). The other expensive component is rooftop solar, but this is the rug that ties the whole room together. We assume you will install enough solar to cover 60–80% of your future electrical load. That solar will be cheap—if we install it the way Australians do and finance it the way we should. That cheap solar energy will be what saves you even more money when it powers your electric household.

## THE FINANCING MODEL

A mortgage is a time machine, which is a concept so important that we'll dedicate chapter 12 to the idea. It allows you to have the future you want tomorrow, today. So if what we want is a future that is safe for our children, a future with a stable climate and no carbon emissions, we need to make that future possible today. The way the US can do this is by making low-interest financing available to everyone.

For the purposes of this thought experiment, I use simple interest-payment calculations based on the full capital cost of the equipment as the principle (zero down!). I assume the 2020 federal mortgage interest rate of 2.9%, and we use the finance period as defined in table 11.2. The only items that we model with a residual price are the car battery and the home battery, for which we assume that at end of life they will have a value equal to the value of their raw materials for recycling, around $40/kWh.

In a moment, I will plug all these numbers in and do the accounting, but first we need one more thing.

## FUTURE ELECTRICITY COSTS

We need a cost for the electricity we will use to power our carbon-free lifestyle. I simply assume the percentage that will be covered by solar will be high and commensurate with the NREL rooftop technical potential studies.[10] For the whole country, this averages out at about 75% of the typical household's load. We model the cost of the rooftop solar portion of that energy at the financed cost of solar ($1/W). This is a cost that we know can be achieved, as Australia is already, in 2021, hitting that target. Financed at 2.9%, that pencils at around 5¢/kWh. Just five pennies. For the balance of the electricity, we assume the current cost of grid electricity (which averages around 14¢/kWh in the US).

Yes, these assumptions are aggressive, but they are not without precedent, and not beyond what we *know* how to do.

## FUTURE HOUSEHOLD COSTS

I plug in all the numbers, and with the power of computers (or hamsters or gremlins or whatever is inside them) I get the answers in figures 11.4 and 11.5. If we do an okay job on cost reduction, every home will save around $1,000 per year, and if we do very well, every home will save

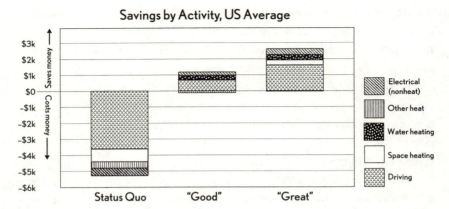

**11.4** If we design our policies around lowering soft costs and develop finance policies to work in partnership with concentrated technological price reduction, all Americans can save a lot of money in the very near future.

$2,500 per year. There are reasons to believe we can do even better than this. What is there to prevent us from financing at an even more aggressive rate if the future of our children is at stake?

If the US aggressively spent research-and-development money on the detailed cost reductions of a penny here and a penny there, we might bring down the cost of the critical components even further. But the prices could stay the same, and with performance improvements we'd also do better. We know we can do solar with an efficiency above 30%, while today it is only 20%. Batteries are the most critical driver of the costs. The cost of battery storage is more about how many times they can be charged and discharged—their cycle life—than it is about their initial cost. Many manufacturers are already extending the life cycle, and hard work should lead to further improvements. If batteries lasted 5,000 cycles and 20 years, instead of 1,000 cycles and 5–10 years, we could exceed even these enticing projections.

## WHAT CAN WE CONCLUDE?

If done right, fixing the climate crisis can save everyone money. If we simply multiply the annual household savings we calculated by the 122 million US households, the country would save $120 billion per year. We

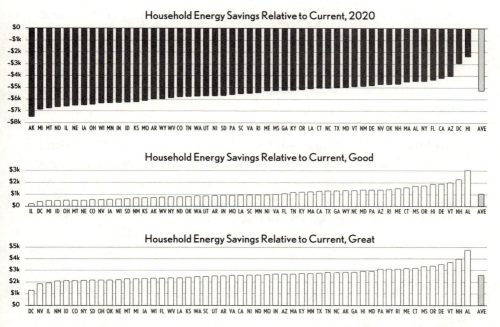

**11.5** Household savings, or decrease in total cost of energy consumed by a household, including transportation fuels and household operational fuels, under the status quo and under three different scenarios of decarbonization aggressiveness.

need to remember the simple mantra: clean electricity is cheaper than fossil equivalents. We can do this, and we can all save money in the process. In the most aggressive model we ran, with batteries at $60/kWh and solar at 80¢/kWh, with an interest rate below 2.9%, we could save more than $300 billion per year. Who said that a Green New Deal had to cost trillions? It will save trillions more.

Up until now, the early markets for clean energy have been developed in places and circumstances that offered glaringly obvious economic benefits. Australia figured out residential rooftop solar because, with the country's low population density, the grid is so spread out that retail-grid electricity is expensive due to distribution costs. South Australia proved out grid-scale batteries because it was cheaper than building out new gas plants. California led the world in electric vehicles because the air pollution in Los Angeles and other urban centers made the need for clean vehicles obvious. In recent years, China scaled this up even further because of even worse air-quality issues. Western Europe and Japan mastered heat pumps because of limited domestic natural gas and the need for inexpensive heat.

If we put together a global recipe of the best of all of these measures, apply massive scales of manufacturing, and eliminate unnecessary regulatory costs, we have a path forward.

During previous emergencies, the first question wasn't, "How can we pay for this?" The first question was, "What do we need to do?" You don't fight a war because you can afford it—you fight a war because you can't afford not to. We can't afford not to fight the war on climate change. We also can't afford not to electrify everything, because if we do it right, it will save us all a huge amount of money.

# 12

## A MORTGAGE IS A TIME MACHINE

---

- ☞ With fossil fuels, you save now and pay later; with renewables, you pay now and save later (including the planet).
- ☞ Most families currently can't afford the up-front costs of decarbonizing their households that will save them money in the long term.
- ☞ If US policymakers can offer "climate loans" at the right rate, the transition to clean energy will start saving us money today.
- ☞ We've created these types of loans before, notably long-term mortgages to enable home ownership after the Great Depression.

As we've seen, clean-energy technologies have higher up-front costs and lower ongoing costs, and the challenge is providing access to the up-front capital. Climate change doesn't care about your household budget or economic circumstances, and unfortunately this means there is currently a disparity between rich and poor in incentives and access to clean energy.

A wealthy household can afford to capture the potential savings from decarbonizing by electrifying everything. They can afford the up-front capital costs of rooftop solar and electric vehicles and hydronic heat pump systems, because they have access to easy credit and home equity loans. At the other end of the spectrum, low-income families need the economic savings of decarbonization but can't afford to pay for the up-front technologies. This is where the equity part of the climate justice conversation needs to focus. Lower-income families would benefit enormously from the lower household costs of a decarbonized, electrified life. The problem is, they very likely don't have access to the capital to pay for it. We can't solve climate change if we don't figure out how to help everyone afford the future.

Whether we can transition to this cheaper future will largely depend on how we finance it. Basically, it comes down to an interest rate. We need to figure out how to help families buy now and pay later. Fortunately, this is something Americans in particular are familiar with.

Recall that switching to all renewables will cost the average US household about $40,000. Prior to COVID-19, 40% of American households didn't have even $400 in the bank for emergency expenses. Very few people have enough cash to pay for a project like this. If they were to pay for it on a credit card, it would be very expensive, because credit card interest rates hover at 15–19%. If we use the common financing options available for solar today, they'll pay around 8% interest. If they could pay for it with a government-backed, low-interest loan at something like mortgage interest rate of 3–4%, it would be affordable for nearly everyone. These may sound like small differences, but consider a solar purchase that is paid for over 20 years. If you could borrow at a mortgage rate of 3.5%, you ultimately pay about double the original price. If you borrow at a common rate of 8%, you pay 4.5 times the purchase price. So don't even think of financing this purchase with your credit card.

As I've said—and it is worth repeating—a mortgage is really a time machine that lets you have the tomorrow you want, today. We want a clean energy future and a livable planet, so let's borrow the money. In David Graeber's enlightening book, *Debt: The First 5,000 Years*, he builds a strong argument that it is through the creation of debt that we actually create money, so what we are really doing is creating the capital to make our climate dreams come true by creating the confidence that we'll save money in the future that repays the debt.

The key to rapid decarbonization will be to create the same kind of public-private partnerships and innovative capital financing strategies that have long underpinned America's economic engine: loans.

We must invent all kinds of low-interest financing options to help consumers afford the capital investments for twenty-first-century decarbonized infrastructure. Green banks are emerging to finance utility-scale infrastructure, but we need to be more audacious. Our climate loans need to be available as retail financial products, so we can all afford the personal infrastructure that builds the climate change solution into our everyday lives. Those of you who aren't homeowners, or never intend to own a home, might complain about the simplicity of this message and the mortgage analogy; I agree that there must also be financial solutions for renters and landlords, and point-of-purchase financing for better appliances. We need greater financial minds than mine (I struggle with the groceries) to iron out all of the details.

America's lifestyle has been built on loans; the car loan and mortgage were both twentieth-century American innovations. America, and indeed the modern world, would be unrecognizable without these financial instruments that help the bulk of the population afford big-ticket capital items.

Creating a climate loan in response to the climate crisis has clear historical precedent. The modern mortgage market was shaped by the federal government's intervention in another time of crisis: the Great Depression. During the Depression, property values plummeted, and about 10% of all homeowners faced foreclosure. The government stepped in during Roosevelt's New Deal, when Congress passed the Home Owners' Loan Act of 1933. That created the Home Owners' Loan Corporation (HOLC) to provide low-interest loans to families at risk of default (for white families, that is; Black families were left out of this deal and, as a result, largely left out of the middle class, so we must be sure that this "climate loan" is available to everyone). As a result, hundreds of thousands of homeowners were able to pay off mortgages, and the program actually turned a slight profit, defying expectations of massive loss of taxpayer money.[1] This program gave rise first to Fannie Mae in 1936 and Freddie Mac in 1968, and created the lowest-cost debt pool and largest capital market the world has ever seen. (The auto loan has a different origin: Henry Ford wouldn't allow his cars to be purchased on debt because of his religious beliefs, and General Motors' Alfred P. Sloan recognized the market opportunity of making cars affordable to the masses by inventing auto financing. This American financing innovation was the precedent for the modern American home loan.)

Under the New Deal, another program offered low-cost federal financing support for electrification. The Electric Home and Farm Authority (EHFA), originally an offshoot of the Tennessee Valley Authority (TVA), helped provide financing for the purchases of electric appliances—refrigerators, ranges, and hot water heaters. Its focus was rural America (especially the Tennessee Valley), and it was part of an effort to expand the domestic market for electricity consumption.

Another New Deal program, the Rural Electrification Program, helped underwrite the installation of basic electricity circuits throughout rural America. The standard installation was a 60 Amp, 230 V fuse panel with circuits for the kitchen and an outlet for a light in each room. Manufacturers that wanted to participate had to produce standard-issue, low-price appliances subject to EHFA approval. Consumers would then select an EHFA-approved appliance and purchase it on an installment credit contract from the dealer, backed by the US Treasury. The terms required the purchaser to put 5–10% down (much lower than any other installment credit offered at the time) and pay down the loan within 36–48 months at 5% interest. The offer was available only to consumers who got their electricity from companies that charged rates acceptable to EHFA. The program ultimately financed some 4.2 million appliances, at a time when there were around 30 million US households.[2]

For the purposes of climate stability and a more robust energy infrastructure, the US government must be just as audacious in financing zero-carbon capital. Tomorrow's infrastructure will necessarily be more personal and distributed, so it's time to help homeowners access the capital they need to contribute to this national effort while also reaping the long-term savings in their homes.

And why shouldn't we finance it like we finance infrastructure? After all, balancing the grid of the future—as we learned in previous chapters—requires leaning on our collective batteries and load-shifting opportunities to make it all work at the lowest cost.

When we electrify everything, everyone will have a personal infrastructure that will not only take energy from the grid, but give some back.

The grand consumer bargain is that the US government should guarantee your cheap loan for your electric cars and your electrified home, in exchange for being able to connect it to the collective national infrastructure, which will balance the loads for everyone.

Clearly, developing financing methods and institutions for this type of infrastructure, including bond measures, public-private financing, and regulated utilities, can significantly aid adoption. Policymakers and manufacturers need to offer solutions with finance, product, and policy, at every one of Americans' purchasing decisions. We also need financing that works for landlords, and for shared infrastructure for people who don't want to own a car or a house. If done right, innovative, low-cost financing will be the most effective way to ensure equity and universal access to cheap, reliable energy in the twenty-first century.

As a result of the COVID pandemic in 2020–2021, interest rates internationally have dropped close to zero. This is the right moment to use these historically low interest rates to finance the household technology and infrastructure that will decarbonize our future lifestyles. Addressing climate change won't work if only the wealthy can switch to clean energy. We must make it possible for everyone to benefit from the savings reaped from electrifying everything—and to collectively meet our climate goals.

# 13

## PAYING FOR THE PAST

- The 8,000-pound carbon gorilla in the room is the proven reserves sitting as assets on the balance sheet of our fossil companies. If we fight these companies until the end, it will indeed be the end, for all of us. With a mechanism for them to fight alongside us, we'll both have a chance to survive.
- Because the stock market was built around fossil fuels, we've incentivized these industries to keep going and we've tied our financial fate to burning fossil fuels.
- Portfolio divestment from fossil-fuel companies is not enough.
- Maybe it will be cheaper to buy out Big Fossil so we can all fight climate change on the same team.

### PAYING FOR PAST SINS

Far back in my family tree were people who introduced coking coal to Australia. My first real job was in the Australian steel industry, which depends on coal. I appreciate my ancestors and the marvels coal has given the world, but it is time to stop using it—for both economic and environmental reasons. Incidentally, I have ancestors on my other parent's side who helped build all of the lighthouses in Ireland, another technology that gave us the modern world, but one that is largely unnecessary now that we have GPS and better maps. The future is here, and the cheapest generation sources are renewables. I have a lot to thank my ancestors for, but we can't afford nostalgia as we figure out how to address climate change.

But as we transition away from fossil fuels, we have to think carefully about the economic ramifications. We've seen how financing can aid the adoption of zero-carbon energy sources, and perhaps we can do the same with fossil fuels.

Digging holes in the ground costs money. Finding the one with oil or gas in it costs more money. Not unlike what I have just suggested for decarbonization technologies, fossil-fuel companies spend a lot to find fossil fuels and only recoup those investments slowly over time. This business model requires borrowing money to dig the holes, and when the companies borrowed that money, the asset they pledged was the oil coming out of their next well.[1]

In the context of the proposed transformation of our energy infrastructure, lingering debts like these are called "stranded assets," and they're a big problem. Stranded assets are resources that once had value but no longer do, usually because of a change in technologies, markets, or social habits.

Currently, it is estimated that the total value of fossil fuels that aren't even dug up yet is maybe $10–100 trillion. We can build this estimate from the 1,500 GT of proven reserves in the ground.[2]

The upper bound of the cost of a proven reserve buyout might be calculated using the price of the most expensive fossil fuel, oil. The price floor of oil is probably the production cost of Saudi oil, which is around $10/barrel or around $60/ton. Our 1,500 GT by that measure is worth around $90 trillion. Most American fields are unprofitable below $30 per barrel. You might be able to buy it all out for the mere value of the profit margin, which is much smaller. The point is, this crazy idea might be a lot cheaper than I estimate here.

Despite the fact that no human has laid eyes on these fossil fuels, they appear as assets on energy companies' ledgers. Climate scientists agree that burning those reserves would compromise the 1.5-degree warming limit—indeed, to stay under that target, we must not burn a third of the oil, half of the gas, and 80% of the coal in that asset pool.[3] Because these fuels are already financed, however, they are already traded like any other form of money. People who own those assets are going to struggle against giving them up. If you had $10 trillion dollars in the bank, would you relinquish it without a fight?

We're living in an economic carbon bubble built on these fuel reserves. If we ban oil and gas companies from extracting these assets, their stocks would crash. That would affect tens of millions of individuals who (perhaps unknowingly) hold these and related stocks in their mutual funds and pension plans. A 2018 study in *Nature Climate Change* estimated that as much as $4 trillion would be wiped from the global economy by stranding fossil fuel assets.[4] By comparison, a loss of only $250 billion triggered the crash of 2008—remember "toxic assets"? Stranding fossil assets would affect not only energy stocks, but also investments in other industries and equipment related to fossil fuel, from gas stations to pipelines to oil tankers. Like the 2008 crash, the rippling effects of such an event could be catastrophic.

Clearly, we can't just pull the rug out from underneath the industry that gave us modernity. We need a plan.

## DIVESTMENT

An activist investment movement known as "portfolio divestment" has been promoted by many liberal-leaning university endowments and is gaining steam. Investment portfolios that join this movement sell off all of their stocks in fossil assets. The idea is that if enough people sell these assets, we'll slowly starve the fossil fuel industry of the precious capital they need to keep digging, drilling, and pumping.

Divestment (also known as disinvestment) can work, and is not without precedent. In the 1980s there was a widespread movement to divest from South African businesses involved in apartheid. In 1986 this divestment campaign was even written into law in the US as the Comprehensive Anti-Apartheid Act. Ronald Reagan tried to veto it, but the Republican-led Senate overrode his veto.[5]

Unfortunately, there are still too many buyers who will purchase these assets from the groups divesting from them. Given enough time, this strategy may work. In no way do I discourage these efforts, but the urgency and inevitability of climate change demands that we move faster and adopt strategies with more guaranteed results. Because divestment is a conflict-based strategy, activists will have to fight every inch of the way, instead of coming up with an amicable solution with wide support.

## STOP FIGHTING; START COLLABORATING?

In navigating this precarious scenario, the best strategy may be to treat the owners of these assets, the fossil-fuel industry, as friends rather than enemies; after all, they did provide us with reliable vehicles and warm homes for a century. Rather than make opponents out of these companies, what if we engage them as the best allies to build the decarbonized future? Today's fossil-energy companies are extremely good at financing capital-intensive businesses. They have enormous teams of smart and competent people who are good with shovels and trucks. They speak infrastructure as a native tongue. Those people could be just as happily— probably more happily—employed building the infrastructure for decarbonization. Why don't we celebrate them as having done an incredible job bringing us the energy we so obviously have enjoyed using? And at that celebratory toast to them for a job well done, let's invite them to be a driving force in our mobilization of decarbonization.

The only roadblock is the stranded assets that keep our friends tied to their old industry. So what if we were to buy them out? It probably wouldn't even be that expensive. We could negotiate. We don't have to buy them out for the full value of their assets, because they would only ever make a slim profit margin (around 6.5%) on them anyway.[6] Let's round it up to 10% to be generous; 10% of $90 trillion is $9 trillion. This is a small fraction of the $100 trillion annual global Gross Domestic Product (GDP). For that price, we could buy back the land and the fossil fuels underneath them (and perhaps even make an international collection of national parks for perpetuity?).

If this were to happen, fossil-fuel companies would wind up with a huge amount of clean capital they could invest in the new energy economy and the new infrastructure of the twenty-first century. Yes, they would have to spend a decade or so winding down their old operations, but they would be optimally positioned to capitalize and operationalize the new energy economy, generating jobs and economic opportunity in the process. Their margins would increase as they built infrastructure spanning supply- and demand-side technologies, and they could leverage the initial capital investment to build businesses with valuations far exceeding that of their stranded assets.

Admittedly, this is a bold idea, but consider it a token of the type of thinking we must embrace to solve our climate crisis and its inherent conflicts. Business as usual will not cut it. You may be an economist or fossil-fuel company executive who is fuming at my naivete right about now, but hopefully it has inspired you to consider a bold idea. This might be the ultimate grand compromise to engage our biggest energy companies in the biggest energy infrastructure build-out ever to occur.

# 14

## REWRITE THE RULES!

- Fighting climate change involves the long, hard, tedious work of changing thousands of regulations.
- Australia is proof that rooftop solar would provide the cheapest energy, if only the US got rid of outdated regulations.
- Building and electrical codes need to be updated to support, rather than conflict with, clean-energy technology.
- We must end all fossil-fuel subsidies.

You may be tempted to skip over this chapter, given the title. It is about boring, bureaucratic regulatory details—but it is vitally important. The lawyers and politicians need jobs, too, so let's get them involved in fixing climate change.

It is not obvious, but one front line in the fight to fix our climate lies with the hundreds of little regulatory barriers preventing the future we need. It would be awfully satisfying if marching in the streets and buying electric vehicles were all we had to do to stop climate change in its tracks. But winning the fight for our future isn't just about marching on City Hall. It's about walking in to talk to your representatives or, better yet, getting yourself elected so that, all across the country, we can make local building regulations, state utility regulations, and federal financing regulations align to support a carbon-free future.

I have always believed that rules and regulations should have expiration dates. Most laws shouldn't last longer than 20 years, because given enough time, humans will figure out how to corrupt or work around any set of rules and regulations. Nowhere is this truer than in the burning of fossil fuels. I'd like to emphasize here that cleaning up our rules and regulations isn't just about new legislation; it's also, critically, about striking down old laws that are broken.

The old way of doing things is embedded in legislation and dinosaur thinking across the country, such as building and electric codes that aren't friendly to solar, home, and vehicle electrification. Similarly, we have backward-looking utility regulations, road rules, gasoline taxes, homeowner-association charters, and tax incentives that all pervert the energy market and prevent us from doing what we need to do. The United States will solve climate change if we don't let the bureaucratic crud and mental laziness of a century of writing regulations for a fossil fuel–based economy get in the way of a green decarbonized future for our children.

## VEHICLES

Australia tried to support its domestic car industry by putting high import taxes and higher luxury vehicle taxes on cars from abroad. Rather than elect to innovate, perhaps in electric vehicles, Australia chose to try to protect its fossil-fueled car industry. Today, it's still expensive to buy an electric vehicle in Australia because of these taxes—a Tesla is twice the price there that it is in the US. Instead of sticking with those regulations, Australia should incentivize its car market to make the cheapest EVs. This strategy has worked in Norway, where electric cars now make up 60% of new car sales and the sale of new fossil-fueled cars is on the track to zero by 2025.[1] Ironically, Australia's policies didn't even save its auto industry; the last Holden Commodore (a red one) rolled off the assembly line in 2017.

In the US, CAFE fuel standards were devised to motivate the American automobile industry to manufacture more fuel-efficient vehicles. That's a great idea. But, as with any set of rules, over time enough lawyers can be thrown at them to find loopholes and workarounds. Light trucks were placed in a different category, with different fuel standards than other vehicles, and because of that, SUVs and cross-over vehicles were born, effectively killing off the market for sedans and shorter, more aerodynamic (and hence more efficient) cars. Efficiency standards are a great idea in theory, but they, too, can be bastardized.

Gas taxes were a reasonable idea to help pay for roads. But America kept them too low for too long. They have been held at the same value, in cents per gallon, since 1993, making the tax proportionally lower and lower every year. This results in badly maintained roads—which many Americans literally feel every day. Bad roads also encourage consumers to buy larger, heavier, gas-guzzling cars. One of the reasons Europe and Asia have smaller, more energy-efficient cars is that they have higher gasoline taxes, which increase the cost of driving. Some people wonder what will happen to these tax revenues when the majority of vehicles are electric. If the US were guided by prudent policies, we would tax vehicles by the mile and by the ton. Something similar already exists with car insurance that charges by the mile. This should encourage lighter, more efficient vehicles that will be driven less. Car companies would be rewarded for lighter-weight vehicles.

In New Zealand, there's tax to pay when a company gives a car to an employee—it's a reward for employment, so it's taxed. Unfortunately, an exception was made for utility vehicles, under the logic that if it's full of tools, then it's not the vehicle you'll use to pick up the kids and go shopping. So all company cars are now "utes" (which is kiwi for truck), whether or not the employee actually needs it, thus evading the fringe-benefit tax. This loophole was only recently repealed, but it is typical of the type of perverse incentives that impact our global energy ecosystem and carbon output.

Even some well-meaning regulations and incentives need to be scrutinized. The early electric-car tax credit of $7,500 was meant to incentivize people to purchase clean-air vehicles and build the electric car industry. Because early EVs were expensive, this looked like a subsidy for the rich. An awful lot of "incentives" are tax deductions or tax breaks. You need to have a pretty high income before you can take full advantage of them. As we move into our decarbonized future, it is worth remembering that we don't win unless we all win, and designing regulations and incentives that work for everyone is critical.

## ROOFTOP SOLAR

As we've seen in the cost difference between rooftop solar in the US and Australia or Mexico, regulations are a serious impediment to widespread rooftop-solar installation. Recall that when you buy solar on your rooftop in Australia, it costs $1/W. In the US, because of regulations, permitting, inspections, and high sales costs, that price is $3/W. The underlying hardware is incredibly cheap, with modules (assemblies of solar cells) selling internationally at 35¢/W (with believable pathways to 25¢/W). Solar energy is not expensive. The regulations surrounding solar make it expensive.

Some of these regulations are so old as to be museum pieces. In San Francisco, you can't put solar modules all the way to the edge of your roof—you have to set them back four feet. I have been told this is because of the fires that followed the 1906 earthquake, which were more damaging than the earthquake itself. It's incredible to think that at that moment in history, the majority of home lighting came from dozens of tiny little fires in your house connected by gas lines. Gaslighting as a climate-change problem has existed for a century! When the earthquake hit, the gas lines leaked, the gas filled the houses and rose to the top because methane is lighter than air. Fires sparked up everywhere.

Subsequently, firemen insisted on building codes that allowed them to vent the building by punching a hole in the roof (one of the reasons the stereotypical fireman carries an axe). San Francisco's lots are small, typically 25 feet wide and 80 feet long. Houses can usually only stretch 45 feet into the lot. The roofs are tiny, and if you eliminate 4 feet around all the edges, you lose 44% of the area that could be used to generate cheap solar electricity.

This origin story may not be exact, but the point is valid: we have building codes all over the country that are in conflict with building the best clean-energy electrical systems. Similarly, our electrical, fire, health, and safety codes; speed limits; environmental laws; and pollution standards were all written for our old fossil-fueled world. We have an opportunity to lower the cost of our new electrical world by sending in an army of lawyers and citizens to clean up and rewrite the codes to optimize a safer, cheaper energy system.

An example of progressive, forward-thinking building regulations are the California[2] and San Francisco[3] requirements for the inclusion of solar PV in new construction. Critically, the California building codes consider the impact on housing affordability—ensuring that the requirement actually decreases the cost of homeownership and passes these savings to residents. But we only build new homes at the rate of about 1% of total US housing a year. We won't solve climate change unless we make the rules, regulations, and incentives apply to upgrading and retrofitting existing homes, too.

Another example that gets a lot of press are natural gas connection bans, which were first applied to newly built homes in Berkeley, California,[4] but are now becoming a national movement after being adopted in Massachusetts. In full disclosure, my friend, an architect named Lisa Cunningham, was instrumental in leading the fight in Massachusetts that sought to remove natural gas lines when undergoing major renovations.[5] Lisa's fight has been contested, which is all the more reason for citizens to take this fight to every other zip code in the country.

## FOSSIL FUELS

In 1913, the first US oil industry subsidy was written into the federal tax code. Called the Revenue Act, this subsidy allowed oil companies to deduct oil in the ground as capital equipment in order to write it off as a tax deduction. It began as a 5% per barrel deduction, and it now stands at 15% per barrel, amounting to billions of dollars annually. This is just one of the many ways the US subsidizes the very thing that is threatening our beautiful world.

A bonding requirement is a deposit that the government requires of oil and gas drillers before they can drill. President Kennedy set these bonds at $10,000, and they haven't been updated in the over 50 years since they were put in place. The bonds are so low that they encourage irresponsible operations, particularly for fracking, where they have led to groundwater contamination.

Must-run contracts are often used by fossil fuel plants to gain monopoly. The fossil-fuel companies argue that they must be allowed to run their coal plants—at the expense of other electricity plants that might be cheaper, like solar. The logic behind this is that otherwise, the coal plants won't be economically viable enough to provide a "reliable grid" when they are needed. I say, let that be so. Let the economics of renewables shut these plants down. Obviously, we are at a threshold where we should provide extra scrutiny of any such contract, as well as any regulation, incentive, tax, subsidy, or rule that advantages fossil fuels.

## ELECTRICAL CODES

National electrical codes are a good idea and are largely written to ensure safe practices. But once again, they were written for a bygone world and for yesterday's technology, not tomorrow's. While national codes need to be conservative, we should push them to embrace the future ever faster. As an example, we currently have codes that require the load center— that's the giant breaker box between the grid and your house—to be sized as though every single load in your house were turned on at the same time. If we electrify everything and triple the load in your house, the peak loads are going to be gigantic, and this quickly goes from a cheap, simple box to a heavy, expensive one. Installing solar as a retrofit already requires nearly half of homes to replace their load center. Given that we know how to make switchable circuits, and since we can manage our peak loads with those switches, we could instead write codes that embrace cheaper switching breakers.

Unions are not guiltless in creating impediments to the future. The electricians' union is apparently largely responsible for the requirement that wiring be housed in a hard conduit—those metal tubes that snake around your basement and on the side of your house. New "soft-conduit" options exist and have been deemed safe in many applications and in other countries. We could embrace new technologies and ways of doing things that would lower our energy costs, too. A decarbonized future will need more forward-looking union practices.

## GRID NEUTRALITY

Maximizing the savings of electrification requires minimizing the cost of the grid, which means grid regulations are critical. I have already mentioned the idea of grid neutrality, where people could share energy democratically, like they do information on the internet. This will not only help with the problem of intermittency, it will also reduce energy costs.

Net metering, where solar panels and other home renewable-energy sources are connected to a public-utility power grid and surplus power is transferred back onto the grid,[6] isn't good enough. Since electricity is generally purchased back at the wholesale rate, rather than the consumer rate, it doesn't encourage you to maximize your own solar capacity or share your storage assets. It's a bit like a tax credit; it's only useful if you pay a lot of tax.

Time-of-use pricing, where electricity rates vary over daily or yearly cycles and utility companies charge more during high demand and less during low demand to help balance the grid,[7] isn't good enough either. This method breaks the day into chunks at different prices, and then consumers choose when to use energy. Not everyone has that choice, and the coarseness of the rate schemes limits adoption.

In a grid-neutrality system, households and utilities would be treated the same and would be allowed to buy and sell from each other without limit. Only through this arbitrage can we realize the most savings (in both dollars and watts). It would be like the internet, where I can give the internet as much information as I want, take as much information as I want, and even create my own businesses.

The utilities don't love this idea, especially those that are also trying to protect their natural-gas business. But remember that "we the people" regulate the utilities, so we don't need to fear them. We can control them; we just need to express our collective will. Utilities will say that they are necessary to provide guaranteed access to low-cost energy to the poorest households. I counter that we can lower the cost of energy to those households if we write the rules of the road correctly. We can guarantee access by other means. The utilities wish to maintain the monopoly that we granted them. If they don't work with us for a climate-friendly future, we should take their monopoly away. Utilities have a big role to play in solving climate change, but that doesn't involve preventing house-holds from generating and sharing electricity for themselves and with each other.

There are thousands of other examples of rules and regulations that undermine the climate action we need today. This is the very front line of the fight we have to save the beautiful world that we want and need. There are good groups working on these regulations, either writing new ones or overturning old ones. (A good example is the Environmental Law Institute of Columbia University and the Widener University Delaware Law School.[8])

There's no such thing as too many people working on fixing these impediments to our future.

# 15

## JOBS, JOBS, JOBS

> ☞ Decarbonizing America in the timeframe required to beat a 2°C/3.6°F increase in global temperature will create tens of millions of jobs.
>
> ☞ High unemployment caused by the coronavirus pandemic presents an opportunity to build a zero-carbon economy with a stimulus effort that can pay for itself.
>
> ☞ The majority of jobs that are created will be distributed throughout the economy, and there will be high-paying jobs in every zip code.

I wish that decarbonizing for the sake of having a better planet to live on would be enough incentive to get it done. But people are rightfully cautious about the impacts this decarbonization might have on the economy. A lot of people have portrayed the idea of decarbonizing America's energy system as being bad for economic growth, particularly for people who work in traditional energy industries. Any proposal to transform the world by overhauling the energy sector needs to reassure people that they won't lose their jobs—or even better, that they will get new jobs that pay more and are more satisfying.

So far, I've outlined a path that can save everyone money tomorrow, but people need jobs today. As I write this, during the COVID-19 pandemic, the unemployment rate is higher than it has been at any time since the Great Depression. There is a solution to this tragic challenge. The good news should be shouted from the rooftops: a rapid transition to a clean-energy economy will create millions of better-paying jobs. In this terrible employment environment, decarbonizing America's energy system is probably the only project ambitious enough to get everyone back to work. These jobs will be highly distributed geographically and difficult to offshore.

## WHY DOES CLEAN ENERGY CREATE MORE JOBS THAN FOSSIL FUELS?

Simply put, clean-energy technologies require more labor in manufacturing, installation, and maintenance than fossil fuel technologies. It takes more people to install and keep a wind farm running than it does to drill a well and keep it pumping to produce the same amount of energy over time. Renewables get their fuels for free, whereas fossil fuels cost money. It takes more labor and maintenance to access those free renewable fuels.

## WHAT DO PEOPLE DO ALL DAY?

In order to have a smooth transition to zero-carbon energy, we have to bring along the people who work in the fossil-fuel industry. But there aren't as many of them as you might guess. The Bureau of Labor Statistics (BLS) maintains excellent publicly available data on jobs in their "Current Employment Statistics" monthly reports. We arrange it in figure 15.1 as a tree map that breaks down the big categories into increasingly small ones—answering the question that Richard Scarry sought to answer in his famous children's book *What Do People Do All Day?*[1]

What stands out is just how few people are directly employed by the energy industry—about 2.7 million of the 150 million (pre-COVID-19) workers in the US. The majority of people employed in fossil fuels are the nearly one million working in gas stations; convenience stores sell 80% of the gas in this country.[2] But we need to remember convenience stores also sell hot dogs, cigarettes, and lottery tickets, so we probably shouldn't categorize them solely as energy industry employees.

# WHAT DO PEOPLE DO ALL DAY?

Energy
1,838,070

| | | | | |
|---|---|---|---|---|
| Education and health services, 24,534,000 | Ambulatory health care services, 7,830,300 | Hospitals, 5,251,400 | Social assistance, 4,224,200 | Educational services, 3,839,200 | Nursing and residential care facilities, 3,389,300 |

Professional and business services, 21,523,000

Professional and technical services, 9,678,800

Administrative and support services, 8,927,600

Management of companies and enterprises, 2,451,000

Leisure and hospitality, 16,808,000

Full-service restaurants, 5,608,200

Limited-service restaurants, 4,572,700

Arts, entertainment, and recreation, 2,480,700

Accommodation, 2,095,400

Other services, 5,935,000

Personal and laundry services, 1,536,000

Repair and maintenance, 1,371,000

Trade, transportation, and utilities, 27,832,000

Electric power distribution, 212,700

Retail trade, 15,669,000

Wholesale trade, 5,937,500

Transportation and warehousing, 5,678,500

Construction, 7,593,000

Building equipment contractors, 2,303,300

Construction of buildings, 1,676,000

Oil and gas pipeline construction,152,400

Manufacturing, 12,844,000

Durable goods, 8,052,000

Nondurable goods, 4,792,000

Mining and oil and gas field machinery, 69,500

Mining and logging, 712,000

Mining, 658,400

Local government, 14,669,000

State government, 5,190,000

Federal, 2,855,000

Government, 22,714,000

Financial activities, 8,823,000

Finance and insurance, 6,475,500

Real estate and rental and leasing, 2,347,400

Information, 2,894,000

Publishing industries, except internet, 766,300

Telecommunications, 706,600

**15.1** All US jobs, prior to the COVID-19 pandemic. Data from the US Bureau of Labor Statistics' "Current Employment Statistics" reports, n.d., https://www.bls.gov/ces/. Get out your glasses!

We can see just how few jobs there are in coal mining—around 50,000—and compare that, say, to the 450,000 people who work in hair styling and barber shops, the 370,000 who work in golf clubs, or the more than 10,000,000 who work in restaurants. There are more accountants in the US than there are employees in the entire energy industry. It's not a big slice of the economy at all.

## HOW MANY NEW JOBS WILL WE HAVE IN OUR CLEAN ENERGY WORLD?

There are many ways to calculate the number of new jobs that will be created by decarbonizing the US, and while estimates vary widely based on methodology, just about everyone agrees the answer is "a lot." My friend Jonathan Koomey warned me that calculating jobs in the energy sector is a fool's errand. I went on that fool's errand in a white paper, "Mobilizing for a Zero-Carbon America: Jobs, Jobs, and More Jobs."[3] I found a new friend, Skip Laitner, an economist used to such calculations, to help me be a fool.

Our estimate of jobs comes from understanding how much energy we currently use in the US, and how much renewable energy we would need to produce to power our lives at the same level of comfort we enjoy today (cars, heaters, push-button conveniences)—all of which I've described in previous chapters. Laitner and I have used this understanding of our energy needs to build a "machines-up" account of decarbonization, counting each specific piece of equipment required to make the transition: solar panels, heat pumps, electric dryers, and electrifying equipment such as hot water heaters and electric vehicles that can be used for energy storage. Then we figured out how many jobs it will take to create all these new electric things.

Economists estimate job creation by starting with a cost estimate. We use our estimate of the cost of all the machines we need to build to figure out how much money the whole project of decarbonization will cost. Economists then draw from historical data the number of jobs created per million dollars spent, for a variety of industries. These jobs include direct, indirect, and induced jobs.

Direct jobs are those that are concretely and specifically in energy. Indirect or supply-chain jobs are associated with servicing the direct jobs.

A direct job might be installing natural-gas pipelines or solar panels, and an indirect job related to that is making the steel for the pipes, fiberglass for wind turbines, or the valves and pumps for the pipeline. Induced jobs are those created in the community around the direct and indirect jobs— the people employed in the restaurants, schools, local retail stores, and other facilities who support the people in the direct and indirect jobs. The woman installing wind farms gets a handsome pay check that she'll spend a good portion of in her local economy employing butchers and bakers and LED makers.

To create our beginning cost estimate, we made a list of what the US will need to build. Remember, we will need something like 1,500 GW of new (clean) electricity capacity on the supply side. That will mean millions of miles of new and upgraded transmission and distribution to get the electricity to the end user. On the demand side, we'll need to electrify our 256 million cars and trucks, 130 million households, 5.5 million commercial buildings covering 90 billion square feet, and all of our manufacturing and industrial processes. From those numbers we can estimate how many batteries, heat pumps, induction stoves, electric cars, and water heaters that will need to be manufactured and installed.

We add up the cost of everything I just described, in comparison to the things they replace. This gives the relative cost of decarbonization versus business as usual. We divide that amount of money by the ratio of direct jobs per million dollars spent for our zero-carbon economy. Similarly, we can calculate the number of indirect and induced jobs. For example, $1,000,000 (2017 dollars; economists have to adjust everything for inflation) spent in construction creates 5.38 direct jobs, 3.87 indirect jobs, and 10.22 induced jobs. That's nearly 20 jobs created per million dollars spent.

That gives us the gross number of new jobs. Then you have to subtract the jobs that will be lost in industries catering to the fossil-fuel economy, including indirect and induced jobs. We have to phase out coal mining and find jobs for those 50,000 miners, but we won't phase out the 2,500,000 jobs in the auto industry, as they'll be redirected to electric vehicles and other net-zero vehicle options.

We assume that we will have a massive wartime mobilization period up front (3–5 years) to get our production capacity up to scale, followed by a 10-year period of deployment. This is in line with an emissions trajectory that increases global temperatures by no more than 2°C/3.6°F. On the demand side, we replace incumbent technology at the rate implied by their natural lifetimes. For example, when your water heater kicks the bucket after 11 years of use, we assume you'll replace it with one powered by a heat pump.

Transitioning to renewables will add a lot of jobs in finance, R&D, and training, which we include.

Figure 15.2 summarizes the output of this model. At its peak, the model projects that this rewiring of America will create more than 25 million new jobs. There are around 12 million jobs currently in the energy industry (including all of the indirect and induced jobs). You can see over the course of 20 years that the existing fossil jobs transition to new clean-energy jobs, and that the end result after the rapid buildup is a sustained 5–6 million job increase over what it is today.

## JOBS PROJECTIONS

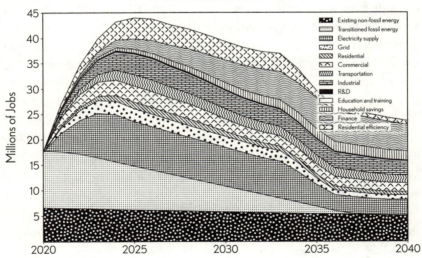

**15.2** Total jobs in the energy sector through 2040, with a decarbonization effort commensurate with a target of 2°C/3.6°F in global temperature increase. The "efficiency" jobs (horizontal wavy stripes) are optional and unnecessary for decarbonization, and they are not included in our total job count.

## WHAT DOES HISTORY HAVE TO SAY ABOUT THIS?

Creating this many jobs, and doing it in quick order with a massive mobilization, is not without precedent. As we've seen, we did something quite similar during World War II. Winning the war for the Allies had a total cost to the economy of around 1.8 times the 1939 GDP. (In 1940, the US GDP was $100 billion. Between 1939 and 1945, the US spent $186 billion producing the war materials critical to the success of the Allies.) Transitioning to a completely decarbonized energy system probably has a cost closer to just one 2019 GDP of $22 trillion—a comparative bargain to save the world.

The last time unemployment was this high, during the Great Depression, we stimulated the economy with the New Deal, which created many jobs but still wasn't enough. Figure 15.3 shows how, at the height of the Great Depression, US unemployment was over 24%. FDR's public-works and jobs programs made real progress starting in 1935, but it wasn't until the war that the job situation changed significantly. After the mobilization of American industry to manufacture war materials, unemployment

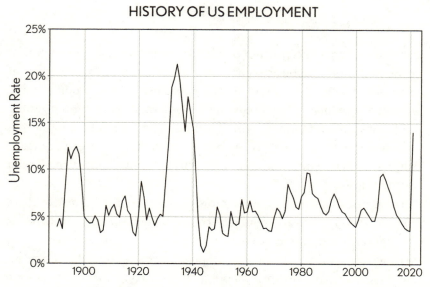

**15.3** Historical rates of unemployment in the US, including the recent COVID-19 spike in unemployment.

was down to 1.2%. Unemployment was so low that, for the first time, women and African Americans were employed in large numbers in high-paying jobs. The productive capacity America built for that wartime effort created not just temporary jobs, but jobs for decades afterward.

We can take a retroactive look at the wartime production known as "the Arsenal of Democracy."[4] Our projections, which look enormous, are not dissimilar in their effect on the economy to what was seen in WWII. There was a 60–70% expansion of manufacturing employment, a more than doubling of manufacturing output, and massive increases in construction and raw-materials production required to feed this activity.

World War II production statistics show the economy-wide benefits of such an audacious project: an 18.3% increase in the labor force, a 63% increase in manufacturing employment, a 52% increase in Gross National Product, and a 58% increase in consumer spending. The war analogy is not perfect, but it will help the public understand that if we shoot for a victory against climate change with a wartime-style mobilization of the nation's industrial productivity, we stand to benefit enormously economically, and in terms of jobs and consumer well-being.

## BUT WAIT A SECOND . . .

Our numbers aren't gospel, and they're almost certainly on the high side. This is so far outside business as usual that it's challenging to arrive at accurate estimates. The historical data of jobs per million people is based on periods in which the economy was fairly normal. What I'm suggesting would be such an enormous stimulus program as to render a lot of that normal econometric data iffy, at best. Nevertheless, you can conclude that there will be a huge number of jobs created—many, many more than we might lose.

The economist's method underscores a sharp conflict in any of these job estimates—you create more jobs by spending more money! This was why the various announcements of the Green New Deals sounded like an ever-increasing race to spend, spend, spend. If you want the biggest headline about jobs, you just spend more money. (If you'd like to re-examine your relationship with money and debt, go read David Graeber.) This is in conflict with making energy cheaper, which should be our other goal. Making energy cheaper means getting efficiencies of scale and lowering the job count required to do every task. Balancing employment and cheap energy is critical, and we need to think bigger as a society about this issue.

Some have proposed ideas like a universal basic income, but we could just consider something we've done before. In the 1950s and 1960s, America went from a majority six-day work week to a five-day work week. The productivity improvements that came from automation after the Industrial Revolution were sufficient to give most Americans more leisure. I don't know a lot of people who want to give up their two-day weekends. So for me there isn't a conflict between creating more jobs and creating cheaper energy. Let's just automate the work, and lower the cost of energy as much as we can, and then make every weekend three days long. Yay for robots!

Another funny aspect of doing this detailed analysis was underscored by calculating the job situation around LED lighting. LEDs are now so cheap and last so long that they save consumers a ton of money. This means that finishing the project of converting much of America's lighting to LEDs will save money, which, to the economist, destroys jobs. Think of the headline "LED lighting destroy jobs—it's un-American!" Except of course, we Americans like our energy cheap.

## HOW MUCH?

The Green New Deal announcements were met with sticker shock, because these vague plans just had a topline number of $20 trillion. They made it sound like a bad, expensive deal for America. It probably does cost about that much, but this amount will be spread out over 15–20 years. This is mostly spending the country was going to do anyway—everyone is going to buy a new car or two in that 20 years, and appliances, and home retrofits, and all of that spending that was going to happen anyway shouldn't be considered an "extra cost."

And in reality, American consumers will save when we transition to a zero-emissions economy. If the country follows something like the recipe this book outlines, it'll save every family up to $2,500 a year. For America's 120 million households, these savings add up to $200–300 billion a year!

The other important point is that the government won't bear all of the cost. If the government uses a mechanism like loan guarantees for this infrastructure, the government doesn't outlay cash; rather, it uses its heft and reputation to give everyone the best possible interest rate. Similarly, the government doesn't have to pay the full cost of every item to make them cost-effective, just enough to tip the market in favor of decarbonized solutions with the right subsidies that are a fraction of the cost of the whole item.

For instance, the current renewable tax credit in the US is set at 26%. If, for argument's sake, we apply this as the government's share of all these costs, it would only amount to about $300 billion per year for the 15 years of the mobilization. This is only a third of our current military budget. Not only that, but America's household and business savings will pretty much cover this cost.

**We need to change the unhealthy narrative that saving the earth is going to cost us money. It won't. If we do it right, we all stand to reap the benefits and save money—and have longer weekends!**

## JOBS EVERYWHERE

The topic of jobs is inherently political. I spoke to a veteran operative in climate politics, appropriately jaded and cynical to prove it, while looking at all of these numbers. He said, "One million future jobs don't have nearly the political currency of the dozens of jobs of one small, loud interest group or union." That's probably true. We won't be able to win every heart and mind.

But to reassure those hearts and minds, remember that this plan doesn't call for immediately shuttering plants and closing all the components of the fossil-fuel economy. Those jobs will transition out at the replacement rate of the machines that are retiring. It'll mean a slow and steady transition into new clean-energy jobs over the next 20 years.

One thing that really matters to people is where jobs are. The nice thing about the plan I outline in this book is that a huge portion of the solution is in your driveway, on your roof, in your basement. These are jobs that can't be offshored to China or Mexico or even done by robots. These are jobs in every zip code in America, and many are biased toward suburban and rural communities. These are also neither boffin (Australian for nerd) jobs in lab coats, nor minimum-wage jobs in restaurants. These are skilled blue- and white-collar jobs, the great majority in the trades—electrical, plumbing, and construction—that will pay well and are rightfully the kind of satisfying jobs where people will go to work in (an electric) pickup truck, feel proud of their day's efforts and contributions to their community, and be part of the larger national project they are building toward: a better, rewired, America.

## RED VS. BLUE ENERGY POLITICS

But knowing there could be more jobs won't necessarily reassure people whose current jobs are in an energy sector that will change. To pretend this problem isn't political is naive. Currently the red states have the majority of the energy jobs. They are scared of losing them, which is a potential reality, and this is trumpeted as the reason not to move toward a clean-energy future. After hurricanes, people in Texas and Louisiana fret about the environmental damage to their magnificent waterways caused by damaged oil and gas facilities, but shortly after the storm, they return to jobs in fossil production that will cause more of the storms that cause their worry.

After the election in 2016, I was moved to look at the political breakdown of the energy landscape. It was one of the more eye-opening things I had done in a while. As you can see in figure 15.4, fossil-fuel production doesn't skew just a little red. It is overwhelmingly—around 85% of all production—located in red states. Among the issues driving voters in those states are their energy jobs.

It is a similarly interesting story if we look at electricity production, which seems fitting in a book titled *Electrify*. But looking closely, we see that the picture is complicated. Red states outproduce blue states in all electricity-generation categories including nuclear and the blue-state darling, renewables. Putting all the solar on our rooftops that we can will still only produce a relatively small fraction of all clean energy, 10–25% of the total supply. So there will need to be an enormous amount of "industrial" clean energy and big renewable installations.

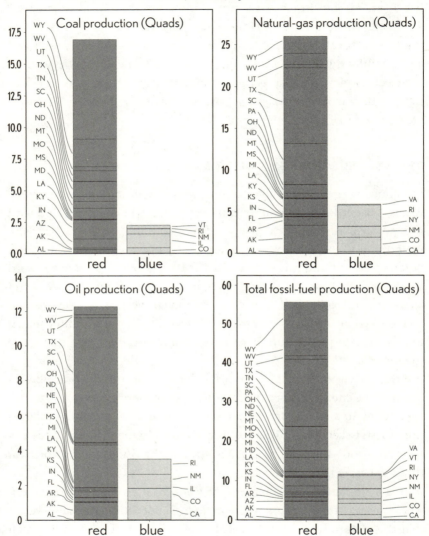

**15.4** Fossil production by state in 2015, including coal, natural gas, oil, and total production, as arranged by 2016 electoral preferences. More than 80% of fossil fuels at that time were produced in Republican-voting states.

# All US Electricity Generation, 2018, by 2016 Election Preference

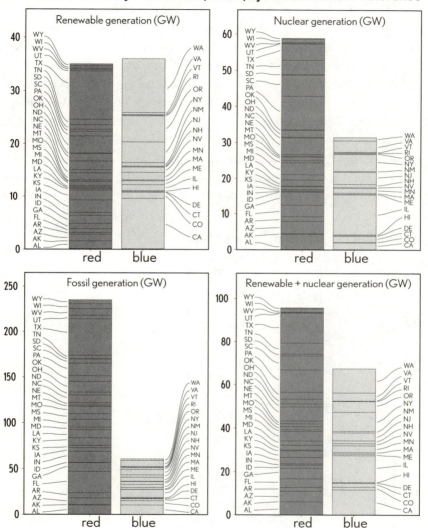

**15.5** Electricity generation by state in 2018, including renewable, nuclear, non-carbon, and fossil fuels, as arranged by 2016 electoral preferences. In all categories Republican-voting states produced more electricity.

Those big solar and wind installations require land. This is the reason more farming, more manufacturing, and more energy production is done in red states: the red states have more land. We can see this clearly in figure 15.6, which shows that around 70% of the US land area voted Republican in 2016. It shouldn't be surprising that that is where the oil is. What also shouldn't be surprising is that the future of clean energy is in those same places. Texas is realizing this with its boom in wind energy installations. There is every reason to believe that the future of energy jobs looks like the past in the critical political sense of employment. Most of the electricity generation jobs will be exactly where the current oil, gas, and coal jobs are for exactly the same reason: these states' wide open spaces and (hopefully soon) clean air.

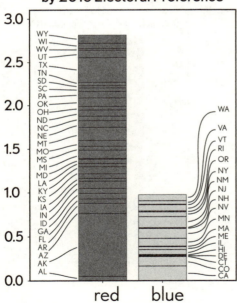

**15.6** Big states vote Republican. The 2016 electoral map, arranged by land area. There is a simple reason for why red states produce more fossil fuels and more electricity: they are much larger. This advantage (around 70% of land area) will also play out in renewables deployment, as renewables require large installations covering a lot of land.

## HISTORICAL PARALLELS

Job creation on this scale (tens of millions) and at this pace (a few urgent years) is not without precedent. The US followed a similar path in mobilizing for World War II (as we'll see in the next chapter). As we have seen, winning the war for the Allies had a total cost of around 1.8 times the 1939 GDP. Transitioning to a completely decarbonized energy system probably has a cost closer to just one 2019 GDP of $22 trillion.

We can look to the wartime production that was recorded in the US War Production Board's October 9, 1945, report, *Wartime Production Achievements and the Reconversion Outlook,* to see that these enormous-looking projections are not dissimilar in their effect on the economy as to what was seen in World War II. In figure 15.7 we see the 60–70%

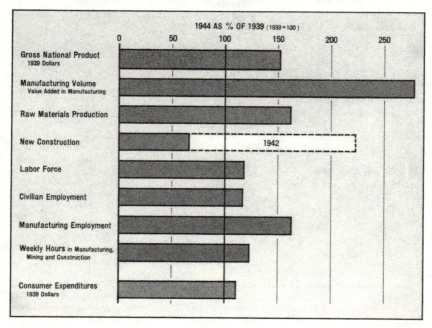

**15.7** Expansion of key US economic sectors comparative to 1939 as a result of wartime production. Source: US War Production Board, *Wartime Production Achievements and the Reconversion Outlook: Report of the Chairman,* October 9, 1945, https://catalog .hathitrust.org/Record/001313077.

# SOME WARTIME SHIFTS IN U. S. ECONOMY

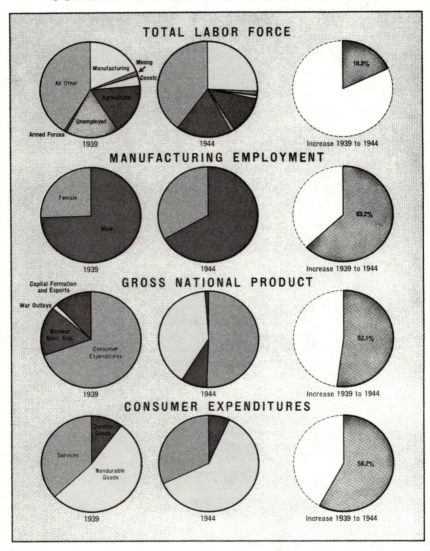

**15.8** Wartime shifts in the US economy for critical economic parameters as a result of the WWII production effort. Source: US War Production Board, *Wartime Production Achievements and the Reconversion Outlook: Report of the Chairman*, October 9, 1945, https://catalog.hathitrust.org/Record/001313077.

expansion of manufacturing employment, the more than doubling of manufacturing output, and the other massive increases in construction and raw-materials production required to feed this activity.

Even more illustrative is figure 15.8, which shows the economy-wide benefits of such an audacious project: an 18.3% increase in the labor force, a 63% increase in manufacturing employment, a 52% increase in Gross National Product, and a massive 58% increase in consumer spending, as so many more people had money in their pockets to spend. The war analogy is not perfect, but it helps us to understand that mobilization of our nation's industrial capacity can drive the creation of millions of new jobs while protecting consumer well-being.

As I'm fond of saying, there will be so many jobs that we'll need robots to do them. Americans need not fear the future if we decide to take matters into our own hands and shape it so that it provides prosperity for everyone.

# 16

## MOBILIZING FOR WORLD WAR ZERO

> ☞ Modern wars are won with technology and a production plan, as proven in World War II.
>
> ☞ Fighting climate change will be cheaper than fighting WWII.
>
> ☞ We need to select a small number of "critical munitions" and ramp up their production rates.

To get to carbon zero, we have to fight World War Zero. (John Kerry may have coined this term, and I use it because it's a great summary of what we need: a wartime effort to get the economy to zero-carbon emissions.) Even if the situation currently seems mired in inertia and political paralysis, we have to act. We must rally together to get to zero-carbon emissions, to prevent a climate disaster that would be as devastating as any world war we could imagine. The odds are stacked against us, but we have a way forward.

As we've seen, electrify everything represents a viable solution that can eliminate most emissions with technologies that already exist. Then the first challenge becomes one of scale: can we produce these solutions in sufficient quantity and within the required time frame? And if not, how quickly can we build the production capacity to make these solutions in quantity?

The fastest possible pathway to decarbonization requires a rapid indus-trial scale-up of working solutions. The US has done this before, with the heroic industrialization effort known as the "Arsenal of Democracy" that rapidly scaled up production to fight World War II. Winning the climate war will take resources and a collective effort akin to what helped win WWII. It will take scaling up industrial efforts at a very fast rate, as we did during that war. As we've seen throughout this book, it's a problem of will, not technology.

Not only did America and the Allies win WWII, we created jobs and technologies that insured our country's long-term prosperity. With a heroic national wartime effort, we clearly can do even better fighting climate change than the already impressive growth rates of our critical clean-energy industries.

In 1939, the United States was at the tail end of the Great Depression. The mood of the country, particularly among New Deal Democrats, was against intervening in international affairs. We see similar sentiments today with the climate crisis—a lack of interest in getting involved, turning away from the problem, focusing on business as usual at home instead of on melting glaciers, rising seas, or wildfires in other places. The climate emergency should be at the top of every politician's agenda, but until after the 2020 election, when President Joe Biden made addressing climate change a priority, it often rated only a token mention.

Similarly, the US was woefully unprepared and disinclined to engage in World War II. In 1939, the US military ranked 18th in the world, just edging out the Netherlands. As Arthur Herman recounts in *Freedom's Forge*, the US Army resources were well behind Hitler's, so much so that Brigadier General George Patton had only 325 tanks to the Germans' more than 2000, and had to order nuts and bolts for them from the Sears and Roebuck catalog. Practice war games held that year were so shabby that the Army used ice cream trucks as stand-ins for tanks, and *Time* magazine reported that the exercises looked like "a few nice boys with BB guns."

In his book, Herman describes how Winston Churchill entreated Roo-sevelt to join the war. After the humiliating 1940 retreat from Dunkirk in a ragtag flotilla of boats marshaled by civilian volunteers, Churchill had to motivate a nation that thought all was lost. His "we will fight them on the beaches" speech was an unvarnished appeal to the fact that it is more noble to go out fighting than to lay down your arms in the face of the enemy. Or as the Churchillian politician we need today might re-write his speech to address our climate crisis:

We shall go on to the end. We shall fight right here in the US, we shall fight for the earth and oceans, we shall fight with growing confidence and growing strength for clean air, we shall defend our planet, whatever the cost may be. We shall fight with our homes, we shall fight with our vehicles, we shall fight with our grid and in the streets, we shall fight in our cities; we shall never surrender.[1]

Roosevelt was convinced, and began a massive buildup for the war effort, employing William Knudsen, a man with a background in car manufacturing, to manage wartime production and gear up industry for the task ahead.

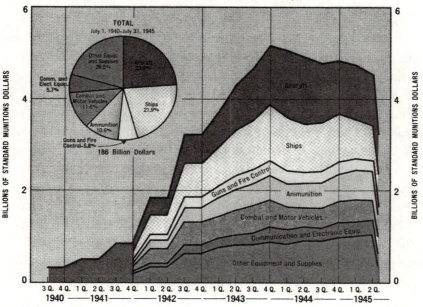

**16.1** The US very quickly ramped up production between 1941 and 1943 for the critical components required to win the war. Source: modified from US War Production Board, *Wartime Production Achievements and the Reconversion Outlook: Report of the Chairman*, October 9, 1945, https://catalog.hathitrust.org/Record/001313077.

The US government drafted a list of critical munitions and offered cost plus a guarantee of a 7% profit to industrialists who would contribute their engineering experience, industrial know-how, and factories to produce a military arsenal that could fight Hitler and save democracy. The profit was sometimes ridiculed as "patriotism plus 7%."

In 1942, Roosevelt appointed another industrialist—Donald M. Nelson, of the Sears catalogue—to the Wartime Production Board. As Roosevelt would say in a speech of that year:

The superiority of the united nations in munitions and ships must be overwhelming—so overwhelming that the Axis nations can never hope to catch up with it. In order to attain this overwhelming superiority the United States must build planes and tanks and guns and ships to the utmost limit of our national capacity. We have the ability and capacity to produce arms not only for our own forces but also for the armies, navies, and air forces fighting on our side. . . .

Only this all-out scale of production will hasten the ultimate all-out victory . . . **Lost ground can always be regained—lost time never**. Speed will save lives; speed will save this Nation which is in peril; speed will save our freedom and civilization.[2]

It did get the job done, and in record time, partly because of the financial incentives the government offered. Building on the mass-production genius of Henry Ford, the Arsenal of Democracy took American-style mass manufacturing to the next level and helped win the war. In 1939 the US had only 1,700 aircraft and no bombers. By 1945, the US had produced a war machine that included some 300,000 military aircraft; 18,500 B-24 bombers, 141 aircraft carriers, 8 battleships, 203 submarines, and 52 million tons of merchant ships; 88,410 tanks and guns, 257,000 artillery pieces, 2.4 million trucks, 2.6 million machine guns, and 807 cruisers, destroyers, and escorts; and 41 billion rounds of ammunition.

It was an arsenal big enough to support the Allies and defeat the Axis. The incredible production ramp up can be seen in the postwar analysis of the project by the Wartime Production Board as seen in figure 16.1.

As none other than Joseph Stalin described it:

I want to tell you, from the Russian point of view, what the President and the United States have done to win the war. The most important things in this war are machines. . . . The United States . . . is a country of machines. Without the use of those machines . . . we would lose this war.[3]

**Table 16.1** Critical Munitions Required to Win World War II Compared to the List of Critical Machines Required to Win World War Zero

| WWII Arsenal of Democracy | World War Zero |
|---|---|
| Aircraft | Wind turbines |
| Liberty ships | Solar farms |
| Bullets | Batteries |
| Combat vehicles | Electric vehicles |
| Engines | Heat pumps |
| Electronics and communications | Grid infrastructure |

In table 16.1, I present an analogous set of "critical war materiels" (as they were known in WWII) for fighting our war on climate change. Instead of air craft, we need wind turbines; instead of liberty ships, we need solar farms; instead of bullets, we need batteries. It won't be simple, and just as with WWII it will require giant political compromises in the public-private partnership required to get it done. Nevertheless, the task ahead can be described fairly simply as a short list of things that need to be deployed at massive scale.

The simple effort of doubling production over and over again is what got the job done during WWII. All of this manufacturing brought over 16 million new people into the workforce. Women, adolescents, retirees, African Americans, and others historically excluded from the workforce were brought in to meet the extraordinary demands of this effort.

No jobs program before or after has been as successful at putting people to work as this US wartime production. After all the smoke had cleared, WWII investments in manufacturing continued to sustain American prosperity for decades. At the height of the Great Depression, US unemployment was over 24%. After nearly a decade of New Deal programs, unemployment stubbornly remained above 14%. With the wartime production effort, though, it rapidly decreased to below what we now think of as the minimum unemployment number of 2%. In 1944, unemployment was 1.2%. Addressing climate change is another project large enough to employ *everyone*.

It may not be realistic to imagine we can keep doubling production capacity until the job is done. Unlike with WWII, there are economic and practical considerations that we now face. What we need to do is double production capacity as fast as possible until we get to the replacement rate, the speed of natural turnover based on the lifetime of a technology. As I mentioned previously, these technologies do have lifetimes. For example, we need to get wind turbine production rates as quickly as possible to the rate at which we can replace them as they retire 30 years after they were built. Globally, if we need 4 TW of wind power and their lifetime is 30 years, then we need to sustainably produce wind turbine generation capacity at 133 GW per year. That is only a little more than two doublings from our current 25 GW, and a production rate we would hit in 2029 if the current industry growth rate of 19% is sustained. If we assume all solar technology lasts 20 years, we need a production rate of 200 GW per year, a rate that we would hit in 2027 if we maintain current growth rates. Once we hit those maintenance levels of production, the industries won't need to grow any more; they just need to continue to produce at that level to sustain the output required for global clean energy. This pushes back by a few years the point when we will cover total global energy demand with renewables, to around 2048 (and around 2045, if we include current nuclear and hydroelectric assets).

The fastest possible pathway is then limited by how comprehensively we adopt these working solutions. As we've seen, this will be limited by their cost and whether US policymakers can implement financing methods that help *everyone* afford the future. A relentless focus on reducing the soft costs—the costs of retrofits, permitting, installation, and inspection that will make this transition easy, cheap, and smooth—is critical, because it is currently not so easy for US consumers to decarbonize. The fastest possible pathway will be enabled by sensible early retirement of our heaviest-emitting technologies, and a sensible regulatory regime that prevents the granting of new fossil leases and exploration rights. This might also be achieved with carbon pricing, although the price needs to get pretty high pretty quickly to actually be the fastest regulatory method. Aggressive research and development is necessary to achieve the fastest possible pathway, but not in the way most people imagine. There is a role for R&D in reducing the cost of the things we need for immediate deployment, but the heavy lift for R&D is in the cleanup projects—finding new ways to decarbonize the sectors that we know we don't have an answer for. Most of these challenges are in either agriculture or materials, and they deserve our attention and resources if they are to provide solutions in the 10–20 year timeframe we need them to.

This wartime mobilization will come with serious up-front costs. But, again, the country has done this before. In 1940, the US population was 132 million, and the GDP was $100 billion. Between 1939 and 1945, the US spent $186 billion producing the war materials critical to the Allies' success, and the GDP doubled from 1940 to 1943.

Today, the US population is 330 million, and the GDP is $21 trillion dollars. If we were to spend in the same proportions today, it would be equivalent to $39 trillion. The good news is that the effort to decarbonize will definitely cost less than $25 trillion—**comparatively less than the financial commitment required to win WWII!**

In fighting World War Zero, the war on climate change, we could completely decarbonize America with a similar effort, on a similar timeline, and at a lower comparative cost to the economy. We know that to meet our climate target rate of 1.5°C/2.7°F, we need a 100% adoption rate, meaning we need every new power plant to be zero carbon, every new car to be electric or zero emission, and every new furnace to be electric and powered by carbon-free sources. That means a radical transformation of American industry and the goods we make.

Roosevelt recognized the necessity of building the military arsenal required to face the enemy earlier than the American public did. When the Japanese bombed Pearl Harbor at the end of 1941, the US military was ready, and the rest of the country woke up to the threat. We face a similarly grave threat today. This is a project that America undertook once before, and we can do it again. In the process, it will reinvigorate our people, pride, and economy.

If we take the massive electrification path to solve climate change—which is really the only viable path—we need to manufacture a very large number of machines. This is not to mention new biofuel industries, new farming methods and technologies, new manufacturing opportunities, and new approaches to forestry.

As we have seen, all this manufacturing will generate jobs for decades. FDR's infrastructure investments of the New Deal markedly reduced unemployment caused by the Great Depression, but the manufacturing effort for WWII was demonstrably more successful in dealing with high unemployment rates. Sixteen million new workers were brought into the economy during WWII. To have a chance of creating the future our children want and need, our war to stop the climate emergency requires a massive buildup of the country's manufacturing, technical, business, scientific, and human resources. We need another Arsenal of Democracy—and we'll need another Apollo moonshot, and maybe a research effort akin to the Manhattan Project as well.

Mid-twentieth-century America was built on an audacious combination of science projects, visionary infrastructure, innovative manufacturing, and novel financing, all supported by and in partnership with the government. This is why the world looks to America to lead the decarbonization revolution: we are the only country with a history of achieving projects this ambitious.

America's abundance has been based on cheap energy, which guarantees economic strength. We can make our energy costs even cheaper, while also meeting the demands of a zero-carbon world. This is the path to a new American abundance. By committing to a full-scale energy transformation through electrification, America will define climate success in the twenty-first century.

Just as America prospered after WWII by building the products that rebuilt the world's destroyed infrastructure, America will prosper after this decarbonization effort by exporting solutions to the rest of the world. We can win this war, and have achieved similar industrial transformations within living memory. As with World War II, we have to fight, and we have to invest right now to save everything we hold precious and dear.

## BATTERIES ARE THE NEW BULLETS

The world produces close to 90 billion bullets every year. That's more than the number of LEGOs produced annually—around 20 billion. What a damning statistic about humanity!

What is a bullet, though, but a metal wrapper around some energy-dense materials? What is a battery but energy-dense materials the size of a bullet, wrapped in a metal canister? We'll need trillions of batteries to power the future if we are using something like the canonical 18650 lithium-ion battery. But if we can make a trillion bullets in a decade, we can surely ramp up battery production.

If we were to increase the production of the small number of things fundamental to winning the war on climate—EVs, heat pumps, solar cells, batteries, wind turbines—what would that look like, and can it be done?

In 2019, the global solar industry added an annually averaged capacity of around 30 GW to bring the total installed capacity of solar to around 127 GW (that is actually generated power, not just nameplate power, which is only realized under ideal conditions). The solar industry is currently growing at 25% annually. In 2018, the global wind industry installed around 20 GW of annually averaged capacity, bringing the installed base to about 249 GW; the wind industry is growing at about 10% per year. In 2019, of the 75 million vehicles sold globally, 1.1 million were electric; the electric-vehicle market is growing at above 20% per year.

As I have shown, a fully electrified economy requires only half the energy the world currently uses. Globally, we use around 16 TW of energy today; half of that is 8 TW—as we've seen, a rough estimate of what we'll need. We should assume some growth in demand, so let's call it a round 10 TW. At the current growth rates, somewhat amazingly, by 2037 wind and solar alone would produce this total energy demand by themselves. At the current electric-vehicle growth rate of 20%, by 2033 we could make the 75 million vehicles we produce globally each year.

This is achieved with the magic of compound growth. If your manufacturing capacity grows at 25% per year, you double your capacity in only three years. This was the logic behind the manufacturing buildup for WWII: identify the critical war materials, and focus relentlessly on increasing the rate at which you can produce those items.

The first liberty ship produced in WWII took 244 days to build. By the middle of the war it took an average of only 42 days. In one heroic publicity stunt, one was produced in under five days.

Imagine for a moment that we got truly ambitious in this climate emergency and doubled the growth rate of production of just these three items: EVs (doubling the current rate would mean a 40% production growth rate), solar panels (50%), and wind turbines (20%). We could meet global demand for energy, with zero carbon, by 2030. All new vehicles could be zero emissions by 2028.

Yes, this is a heroic plan. But in a story where you're saving the love of your life—planet earth—it's worth it for all of us to be heroes.

# 17

## CLIMATE CHANGE ISN'T EVERYTHING

---

> ☞ Electrifying everything is the immediate path out of our climate emergency.
> ☞ But our environmental problems are bigger than climate change.
> ☞ We must rethink industry for a decarbonized world.
> ☞ We could address climate change and still kill the oceans with plastic.
> ☞ We need to think about our products as heirlooms, and recyclable ones, if we are to solve our consumption problem.

Solving climate change isn't much good if we suffocate the oceans with plastics, kill our bees with pesticides, and continue to pollute the world's waterways with excess fertilizers and other environmental toxins. The industrial ecosystem is where our climate-change challenges collide with all of our other environmental problems. There are huge opportunities for win-wins that address not only climate problems but also the other negative effects of our consumption habits.

My original degree was in materials science and metallurgy, and my first industrial jobs were in aluminum smelters, steel blast furnaces, and rolling mills. Apart from apathy, there is no reason to believe we can't massively reduce the energy use of industries like these while also fixing a huge number of other environmental problems associated with how we make things. Rethinking industry for a decarbonized world is one of the most exciting challenges for the industrialists coming of age today.

The US industrial economy is actually the largest consumer of energy (~32%) and a huge emitter of $CO_2$ and other climate-warming gases. We saw the energy flow breakdowns of this sector in figure 4.5. The industrial sector, as defined by the agencies who measure our energy uses and carbon emissions, includes mining, construction, and agriculture, as well as manufacturing, the largest component of the industrial economy. To take one example, nearly one of the 32 quads required to power industry is used for making fertilizer. Fertilizer is good, and we need it, but we don't use it very effectively, and it is likely that a lot could be saved while maintaining healthier and better food systems and soils. Heavy fertilizer application and poor soil management leads to huge emissions of nitrous oxide, a more damaging greenhouse gas than carbon. There are similar opportunities for improvement throughout the industrial sector.

While we can easily understand why our cars produce $CO_2$ emissions from the giant amounts of gasoline we feed them, and why heating our homes and running our stoves do too, it is harder to understand how the things that we purchase as "consumer goods" contribute to emissions.

We saw earlier, in the Sankey diagram in figure 4.5, just how much energy is used in making all of our stuff. This diagram is largely built on the data provided by the semi-annual Manufacturing and Energy Consumption Survey (MECS). I have dived deep down this rabbit hole at various times, as incredible business and research opportunities are buried in the decarbonization of the industrial sector.

A helpful way to understand the flow of energy through the industrial sector is to look at the flow of materials through the economy. We see in figure 17.1 just how much stuff we move around. The 6,544 tons of stuff the US takes from the natural world each year is 20 tons per person. Funnily enough, this is without even counting the $CO_2$. When we burn those 1,936 million tons of fossil fuels they mix with oxygen to create $CO_2$—around 6,700 million tons of the stuff. If we counted $CO_2$ as one of the things we manufactured, it would, astoundingly, weigh more than everything else we push around combined!

Contemplate that before you get too enamored with the propaganda that is the hype around carbon sequestration. We would have to bury more $CO_2$ every year than all of the other stuff we dig out of the ground or take from forest and field. That's going to be one hell of an environmentally

# US Material Flows
## Millions of Tons per Year

$CO_2 : 4912$

oxygen : 2976

Atmospheric contribution

----------------------------------------

Domestic extraction : 6544

petroleum : 594

**fossil fuels :
1936**

natural gas : 552

coal : 790

- - - - crude oil : 486
- - - - natural gas liquids : 108
- - - - natural gas : 552
- - - - other sub-bituminous coal : 380
- - - - other bituminous coal : 282
- - - - lignite (brown coal) : 63
- - - - coking coal : 63
- - - - anthracite : 2
- - - - peat : 0.4

**biomass :
1710**

crops : 738

crop residues : 525

grazed biomass
and fodder crops : 221

wood : 220

wild catch and harvest : 5

cereals n.e.c. : 399
oil-bearing crops : 131
wheat : 56
sugar crops : 54
vegetables : 30
fruits : 28
roots and tubers : 21
rice : 10
nuts : 3
pulses : 3
fibers : 3
tobacco : 0.3
spice beverage pharmaceutical crops : 0.03
straw : 390
other crop residues : 134
grazed biomass : 221
timber (industrial roundwood) : 192
wood fuel and other extraction : 28
wild fish catch : 5

**metal ores :
593**

non-ferrous ores : 536

ferrous ores : 57

copper ores : 358
gold ores : 169
zinc ores : 5
lead ores : 2
uranium ores : 1
platinum group metal ores : 0.6
other metal ores : 0.5
nickel ores : 0.1
bauxite and other aluminium ores : 0.09
titanium ores : 0.03
silver ores : .0001
iron ores : 57

**non-metallic
minerals :
2304**

non-metallic minerals
construction : 2084

non-metallic minerals
industrial : 220

sand gravel and crushed rock for construction : 1088

limestone : 695

ornamental or building stone : 238
structural clays : 23
dolomite : 23
gypsum : 16

industrial sand and gravel : 120
salt : 44
fertilizer minerals n.e.c. : 27
industrial minerals n.e.c. : 15
specialty clays : 13
chemical minerals n.e.c. : 2

**17.1**   US flows (by tonnage) of materials through the economy.

destructive process and require a second industrial ecosystem as big as all of our current ones.

On the bright side, looking at these giant material flows gives us the opportunity to contemplate a saner version of carbon sequestration. Looking at those flows, especially the bigger ones, ask yourself, "Can I conceivably bury or sequester carbon in that flow?" If you can figure out how to make the answer to that "yes," you have an enormous contribution to make in addressing climate change. If we are to store carbon somewhere, at the scale required, it would need to be absorbed in these large existing material flows—like moving soil, or in forestry and wood products, or in concrete and drywall (I helped found a robotics company that can finish drywall, and we are looking at how to use the process to make walls a carbon sink, not a source). It may not be as glamorous as "direct air carbon capture," but it is more feasible and more reasonable. It does realistically mean that our pathways for sequestration at scale are likely slower than have been modeled into UN IPCC emissions reductions scenarios. That means we need to figure it out quickly and immediately get going.

There are many other efficiency wins and energy reductions through technology transformations that can substantively impact industrial energy flows. In addition to using our material economy to sequester carbon, we need to start thinking big about how to use fewer materials to achieve the same goals, how to achieve 100% recycling rates of those materials, and how to use materials that have lower toxicity. Shockingly, one-third of the world's children already have poisonous levels of lead in their blood.[1]

## AN IMPORTANT WAY TO THINK ABOUT ENERGY IN STUFF: EMBODIED ENERGY

Engineers think about the energy or carbon footprint of products in terms of their embodied energy or embodied carbon. Embodied energy—the sum of all the energy required to produce something, which we can think of as energy incorporated or "embodied" in the thing itself—is pretty easy to understand, which is why I use it as the reference number when making calculations. You can imagine that the embodied carbon could

**Table 17.1** Approximate Embodied Energies and Embodied Carbon for an Array of Common Materials.

| Material | MJ/kg | Carbon (kgCO$_2$/kg) |
|---|---|---|
| Concrete | 1.11 | 0.159 |
| Steel | 20.1 | 1.37 |
| Stainless steel | 56.7 | 6.15 |
| Timber | 8.5 | 0.46 |
| Glue-laminated timber | 12 | 0.87 |
| Glass-fiber insulation | 28 | 1.35 |
| Aluminum | 155 | 8.24 |
| Bitumen | 51 | 0.4 |
| Plywood | 15 | 1.07 |
| Glass | 15 | 0.85 |
| PVC | 77.2 | 2.41 |
| Copper | 42 | 2.6 |
| Lead | 25.21 | 1.57 |

vary greatly depending on the energy source that was used to produce the material. If we were making all of these materials with zero-carbon electricity, most would have near-zero embodied carbon. It is embodied energy we care about, for which I share reasonable reference numbers in table 17.1.

But this assumes the material is only used once. In reality, to compare all of these materials, we need to recognize that the energy or carbon impact of an object is determined by the equation:

$$\textit{Energy per utility of a thing} = \frac{\textit{Weight of thing x embodied energy}}{\textit{Lifetime or number of uses}} \quad (17.1)$$

This equation tells you some really important things. To decrease environmental impact, you can lower the weight of a thing, or you can use a different material altogether. But making the thing last longer is key.

The first strategy is the material-optimization strategy used by many companies—for example, shaving a few grams of plastic off your toothbrush. The savings are generally really small from this type of effort. As a strategy, it's similar to energy efficiency: often hyped, yet unlikely to yield transformative gains. Often, though, designers will use more exotic materials to achieve those weight savings, but at the expense of increasing the embodied energy of the object produced. This is generally typified by the use of carbon fibers and exotic composites. These materials end up making lighter objects, but the weight loss is offset by the embodied energy gain by using the new materials, and often by some toxic future recyclability problem, as well.

The other strategy that lots of "green" companies use is material substitution. This is evident in all of the "green" products made of bamboo. People associate bamboo with "green," but it isn't always so. Many bamboo products are shipped from China and grown under unclear environmental standards; "sustainable" bamboo clothing and fabrics are often processed with toxic chemicals, or using enormous amounts of water and heat. The much-touted advantages of hemp-based textiles, perhaps promoted by those smoking their own T-shirts, is eliminated by the unfortunate fact that it takes much more water and heat to separate out the hemp fibers than it does with the cotton production process.

In most cases, it is the number of uses, or the lifetime, of a product that determines its sustainability. If your bamboo toothbrush is only used once, it's an awful idea. If you use your carbon-fiber bicycle for 15 years and 60,000 miles, it was an excellent choice.

In estimating our future economy-wide electrification needs, we don't assume any other efficiency wins in manufacturing, although there will be many. One way to reduce the need for electricity is if we simply buy and throw away a lot less stuff. The great majority of all of the materials we use eventually wind up in a landfill, and landfills themselves are a significant source of emissions as the buried cellulose decomposes anaerobically into methane (which could be used to power a generator). Americans throw out 4.5 pounds of stuff per day, a number that is still increasing. (And remember, that is only your personal stuff that goes to landfill; if we added all of the roads and bridges and your share of the shopping malls and movie theaters you use, it is more like 100 pounds a day!) Even more astonishingly, every American uses an average of 6 tons of fossil fuel per year, which amounts to around 40 pounds per day of carbon, or more than 100 pounds per day of $CO_2$. As we are all learning, there is no "away" to throw it all.

The energy used to make an object is amortized over its lifetime. This is why single-use plastics are a terrible idea. It is also why the easiest way to make something "greener" is to make it last longer. I've always loved the idea that we could turn our consumer culture into an heirloom culture. In an heirloom culture, we would help people buy better things that would last longer, and consequently use less material and energy. There is an old adage that rich people can't afford to buy cheap things of low quality—a statement of the fact that well-made things last longer. This is the environmental embodiment of that parable. Buy good-quality things and use them for a long time. This once again gets to the financing issue, however. Often the right choice is more expensive up front. Again, US policymakers will need to think about how to help consumers finance the right material and product choices.

Vehicles are often the focus of technocrats' obsession with embodied energy, and for good reason. Approximately half of the carbon emissions of a typical car are in the production stage—in their embodied energy. One thing that excites me about electric cars is that they are so simple they should last much longer. If your classic electric car (I can only hope that one day there is such a thing) drives for 50 years with only a (recycled) battery pack change-over, that will eliminate a huge amount of the true energy used in driving.

It is estimated that the manufacturing of an ICE automobile requires about 125 GJ of energy, and because it is a little heavier and the batteries are more complicated to produce, 200 GJ for an EV.[2] That's 50–60,000 kWh. If you were driving that EV at a pretty efficient 300 Wh/mile, that means you have to drive the car for 200,000 miles before the energy used in moving it would equal the energy used in making it. This means it doesn't really get 300 Wh/mile, but instead 600 Wh/mile. (The same math applies to ICE vehicles.) If you multiply the embodied energy in the 17 million ICE vehicles sold in the US every year by that 125 GJ, it pencils out (and sorry for switching units) at a little over 2 quads. Two percent of our current energy consumption is in making vehicles!

Designing vehicles and building them to last 500,000 miles is obviously a much better idea. During the electric-scooter craze of 2018, I calculated the embodied energy of a typical scooter, and used its average 45–day life in service to estimate its total energy consumption per mile. It was close to 900 Wh/mile. That is worse than the fuel economy of a Ford Expedition SUV!

Industrial energy use and material resource use are such important topics that the DOE publishes fantastic studies on just how good we could get at producing various industrial materials. These are known as Energy Bandwidth Studies.[3] They are worth taking a look at, if only to see the biggest energy consumers and carbon emitters. I will provide a look at a few below.

## STEEL

The carbon emitted during the production of steel is a result of the energy used in heating and processing the steel, and of the coal used in the process of making the raw iron in the first place. All of our steels have significant quantities of carbon in them; in fact, carbon content is one of the principal descriptors of types of steel. "Low-carbon steels" are ductile and pretty strong, "high-carbon steels" are typically more brittle but super strong. Today most of the heat for the steelmaking process comes from natural gas, but there is no reason why it cannot come from clean electricity. There are companies all over the world working on ways to add carbon content to steel without having to add it as coal in the blast furnace, where it oxidizes and produces $CO_2$ as part of the process. A Rearden metal–sized fortune (excuse my *Atlas Shrugged* reference) will go to whomever succeeds.

Thyssenkrupp has figured out how to make steel using hydrogen instead of coking coal. One of the critical places for hydrogen in an electrified world is in manufacturing to generate the high temperatures we need for industrial processes.

Steel is 100% recyclable, but only about two-thirds is actually recycled.

## CONCRETE

Cement is another big energy consumer and $CO_2$ producer, but we don't yet have a scalable alternative. This represents a giant opportunity:

Roman and Greek cement absorbed $CO_2$, and this might be a case of back to the future by using cement to sink carbon. Cement is also what we need to make humanity's favorite material, concrete.

I've always been astounded by the statistics on concrete. The US produces almost two tons of concrete per person per year! The stuff is everywhere you look. Joni Mitchell was bang on target when she sang of paving over paradise. It is estimated that 8% of global emissions come from cement alone. Half of those emissions are from the energy required during production, and the other half are emitted in the creation of clinker, the lime-based binder that holds it all together. Limestone ($CaCO_3$) is heated to become lime—calcium oxide ($CaO$)—which creates leftover $CO_2$.

But it doesn't have to be this way. We should be able to make cement that absorbs $CO_2$ through its lifetime. And we certainly should be able to build with less concrete. Covering ground with concrete has negative effects on drainage, soils, and more. I'm sure we can do better.

The American Concrete Pavement Association publishes data about how roads made of hard concrete, instead of softer asphalt, enable cars to get better mileage. The real problem is that we don't add the embodied energy of making the roads that our cars drive on into our estimates of energy used per vehicle mile traveled. This means that our 600 Wh/mile estimate just got worse.

What's more, very little concrete is recycled, though some of it becomes the base for yet more roads.

## ALUMINUM

Most aluminum is already made with electricity, so once again, in theory we can make it without carbon emissions, but the energy input is not the only source of carbon emissions. In the arc furnaces used to smelt aluminum in today, the electrodes are carbon, which is the source of much of their carbon emissions. Apple recently worked with Alcoa and Rio Tinto to create the first batch of carbon-free aluminum.[4] I've personally always found aluminum to be a wonder material, so I'm glad we are on a good track for carbon-free aluminum.

Aluminum is also 100% recyclable, yet only about two-thirds is recycled in the US.

## PAPER

In theory, paper can be a zero-net-carbon product. The huge amount of energy used in the paper and pulp industry (more than two quads!) is mostly used in separating the cellulose fibers from the lignin glue that holds trees together. The promise of the paperless office has yet to lower the amount of paper we need, and the convenience of online shopping is increasing the demand for cardboard packaging. It might not sound sexy, but a better paper and paperboard industry is critical to addressing our climate issues.

About 63% of paper and paperboard are recycled in the US.

## TIMBER

Wood is good. I like to think that wood is the second-best method for carbon sequestration, other than books! To use more would mean we need to have much better forestry management. People are already building wooden multistory housing, and wood is really a perfect sustainable building material, but there isn't enough of it in the world for everyone to have an American-style home. I once planted 30,000 trees—I told you I had an environmentalist mother—when my mom was trying to create habitats for some endangered birds. Many of these trees reached maturity. Just a few of them could have been my entire lifetime's supply of construction materials and carbon sequestration. More forests, better-managed forests, and more wood products that are made to last longer would be a great thing.

## GLASS

Glass can be recycled basically infinitely, but it does require a lot of energy to produce it. This is because the melting point of glasses is so high.

We are getting good at making stronger, thinner, tougher glasses, but maybe all we really need is a cultural shift back toward reusable glass packaging. It's cleaner and chemically much safer than storing your food in plastic.

Every time I receive one of those ubiquitous plastic takeout food trays I think about how 50 years ago it would have been touted as a food-storage miracle. Remember Tupperware? Two hundred years ago it would have been prized as highly as we value gold, pewter, and silver. Now, they get used once and thrown away. There is no "away," though, so let's start thinking about reusable glass.

But glass isn't always the answer; in bottles it is hugely heavy and usually single use. If you are going to drink wine, buy it by the cask. We used to have cultural practices around "house wine" that are worth reprising—many Italians still buy, or make, their wine that way. If you can't get behind a keg or a cask in your basement, then at least consider aluminum cans of your currently glass-bottled beverages, since they are more recyclable and much lighter.

Today, only 34% of glass is recycled in the US.

## PLASTICS

I don't write much in this book about the problem of ocean plastics, nor of the larger plastic pollution nightmare. But it is an enormous concern. Perhaps not surprisingly, the fossil-fuel industry expanded from the low-margin industry of energy supply into the higher-margin industry of plastics. They have had astounding success with the project, as evidenced by the plastics that pervade our marine environments. We need a combination of consumer behavior change and new technology to solve this problem. I have hope for biologically derived plastics that are biodegradable. This is a critically important problem to solve, as the current production of the plastics we use every day generates large quantities of nitrous oxides and other gases even more harmful to our atmosphere than $CO_2$.[5]

Unless we change things quickly, plastics, on their own, will emit 10–13% of our remaining carbon budget.[6] This isn't as obvious as you might think.

Plastics have big, long carbon backbones and last forever, so you would think we could sequester carbon in them. Maybe we should drill for oil to mold giant plastic dinosaurs that we could re-bury to sequester carbon! Unfortunately, this is what most people's carbon sequestration plans are anyway. What actually happens is that in the creation of olefins—the precursor to most plastics—there are large amounts of nitrous oxide emissions.

The recycling picture for plastics isn't great: less than 10% of plastic is recycled in the US.

So recycling doesn't work, and we need an entirely new pathway to plastics, but even if we did they would still gather in the oceans. Because of all this, I think we should use paper and glass and metal and more reusable containers, but we should also invest heavily in synthetic biology or other pathways to a new kind of polymer that would quickly biodegrade the way leaves do. Leaves, after all, don't end up as ocean microplastics.

If the US acts wisely, our moonshot science investment would involve studying and inventing materials systems, especially polymers, that don't degrade our environment or use excess energy. Biology and the natural world has a lot to teach us in this domain.

## MATERIAL PROCESSING

It costs enormous amounts of energy to process raw materials into the objects we use. When I studied metallurgy, we called it "heat and beat," or the study of getting things hot and shaping them with something like a hammer. If you could imagine a better way to do any of these things—grinding of materials (0.49 quads), electrochemical processing (0.16 quads), or food processing (1.11 quads)—you could be a captain of those new industries.

While most of the heat in our homes is low temperature—hot water and hot air at below boiling point—a huge amount of energy used in industry is in high-temperature heat. This is the heat used to bend steel, melt aluminum, and bake ceramics. This high-temperature heat isn't amenable to the efficiency tricks of heat pumps, but relies on looking for other pathways to energy efficiency. There are a lot of opportunities to build billion-dollar manufacturing businesses in the electrified future, if you can figure out the right technology to avoid those high temperatures or make them cleanly.

## THE MATERIALS OF OUR ZERO-CARBON FUTURE

Many renewable technologies rely on rare-earth metals such as neodymium, scandium, and ytterbium for critical components. The rare-earth metals used in high-energy magnets and electronics are actually not as rare as their name implies. Their costs pose some challenges to critical components like electric motors and batteries, and so finding ways to decrease the amount needed can reduce the costs of these devices.

Developing robust and efficient recycling pathways for solar cells, batteries, motors, and carbon fiber will offer further opportunities to lower costs of critical components by lowering material costs.

In this book, I have explained that we need about 4,000 W of constant energy supply per person to live our decarbonized life. That translates to a 20,000-watt solar array. A 400 W solar module weighs around 40 pounds. That's 10 watts per pound. That means *every one of us* needs a 2,000-pound solar array. That array will last around 20 years. That means we will need around 100 pounds or 40 kg of solar arrays built for every one of us every year. We will probably figure out how to make these arrays last longer and work more efficiently, and how to make them thinner and lighter, but even if it is 10 or 20 kg per person per year, that's going to be a lot of what amounts to e-waste.

Similarly, a family of four on the decarbonization program presented in this book will need around 200 kWh of batteries to make it all work. At current battery efficiencies and a lifetime of 7 years, that will be 30 kg or 70 pounds per year of batteries for each of us. Obviously, we need to consider making solar arrays, wind turbines, and batteries last longer. We need to figure out how to recycle *100%* of the materials in them. You could imagine tying the federal financing suggested in earlier chapters to a federal recycling rebate (like the 5–10 cents per bottle applied to certain bottles in some places) to motivate collection and reuse.

Much of the cobalt in the world is mined in West Africa, in Zambia and the Democratic Republic of Congo (DRC). Cobalt is a critical component in batteries and other electronics. I visited my dear friend Louise Leakey, whom I should write to more often and who is the latest in the line of very famous Leakeys who not only study the origins of evolution in humanity's birthplace in Kenya's Rift Valley but are also vital protectors of Africa's incredible wildlife. Louise's father, Richard, lost both of his legs when his plane was sabotaged because he was leading the campaign against ivory exploitation that was endangering elephants. Louise's husband, Emmanuel de Merode, a Belgian prince, has devoted his life to protecting Virunga National Park in the DRC, including its highly endangered population of mountain gorillas. Emmanuel was shot multiple times and (like everyone in the Leakey orbit, it seems) lived to tell the tale, while trying to prevent the encroachment into these precious habitats of the very mining companies that give us our cobalt. If I had to have a world without gorillas or a world without electric cars, I would choose to kill off the EVs and keep the gorillas.

The story of neodymium (which makes the powerful magnets used in computers, cell phones, medical equipment, motors, wind turbines, and other electronics) and the other rare-earth metals is only marginally better than the story of cobalt. After a half century of mining these metals, China has woken up to the enormously toxic trail they leave, not to mention the questionable labor practices involved.

This is to emphasize that we need to be much better at mining, and we need to be enormously better at recycling. We should also spend more of our country's science budget on developing alternatives to these exotic materials, or designing electrical machinery around less-exotic, and less-toxic, materials.

But even the less-exotic materials can be enormous problems. I grew up in Australia during the period when Australian mining companies were pillaging the environment in the exotic rainforests of Papua New Guinea. Whole mountainsides were destroyed to bring us the copper we use in electronics and wires.

It doesn't have to be this way. I have a number of great friends pioneering the use of biology and biological processes to make the things we need less toxic. My friend Drew Endy and a former professor of mine at MIT, Tom Knight, pioneered an area called synthetic biology. It uses the power of cellular manufacturing to produce biological materials. To date, it has mostly been harnessed in pursuit of better pharmaceuticals, but I believe their bigger promise is in solving some of our bulk-materials problems. I took classes with Professor Shuguang Zhang, also at MIT, where we brainstormed how to create materials with the incredible properties of bones, bamboo, fingernails, and silk, but in quantities and formats amenable to a clean, green industrial manufacturing infrastructure. When I tell people I want to make surfboards out of fingernails, they imagine I'm going to do a lot of toenail clipping, but what I really mean is growing surfboards in vats of organic materials. Endy now advocates that mushrooms, or more specifically the mycelium fibers they produce, can be used to make packaging, insulation, and building materials. Our mutual friend Philip Ross is making leather substitutes with it that are fantastic.

My friend Fio Omenetto ran the silk lab at Tufts. He and I brainstorm all sorts of things to make with biology. His favorite new plant is *Lunaria annua*, a marvelous plant with reflective properties that also grows prolifically, like a weed. He imagines using it in geoengineering schemes to change the earth's albedo. (Albedo is a measure of the reflectivity of a surface, like snow or dirt. If more of the earth were covered in snow, more light would be reflected into space, and we would slow or reverse global warming—this is why losing the glaciers and Arctic ice is a terrible idea, because reflective snow is replaced with light-absorbing water or stone.) I am trying to convince Omenetto that we should start making glitter with *Lunaria annua*. Glitter today is a toxic little timebomb made out of plastics and sometimes thin layers of metal (for the sparkles). So for all the wedding guests and children out there who love sparkles but also want sparkly clean oceans, we should be harnessing the powers of this miraculous little plant to make biodegradable sparkly glitter. I would like to call it "Mermaid Glitter—sparkles you can have without choking the fish." (This is why they assign me to the lab and not to marketing. I'll ask my seven-year-old daughter what she'd like to name it.)

The environment, the elephants, the gorillas, the fish, the mermaids—they are all worth saving. These are the things that motivated me to get into the business of finding solutions to climate change in the first place. These are the things that make the world rich, beautiful, and fascinating.

We might be compelled to fix climate change to save humanity, or to save the creatures, or both, but it won't be much of a victory if we lose the apes and the dolphins and the polar bears along the way.

My colleagues in labs and universities and companies around the world are all confident that we can collectively find the answers and build the electrified world my seven-year-old and eleven-year-old deserve.

**It will take all of us.**

# APPENDIXES: RABBIT HOLES

# A

## YES, AND . . .

I want readers to be able to understand the main arguments of this book without getting stuck in too many details. Here, I will try to offer you dinner party–ready talking points for the main questions that people will inevitably have for the main argument of the book. Each topic is worthy of a book in itself. If I dispose of a favorite baby of yours too quickly here, or you think I have it all ass-backward, then we should grab a beer sometime.

### YES, AND . . . WHAT ABOUT CARBON SEQUESTRATION?

Carbon sequestration would be a great technology to support, if only it were a good idea. It is attractive because it gives us the illusion we can just keep on burning fossil fuels if we can figure out how to suck the emissions back out of the air.

This idea derives from the natural processes that have kept our planet in balance for millions of years. Trees, plants, and microbes evolved to turn atmospheric $CO_2$ into a useful product—biomass or wood. They do so using cascades of elegant chemical reactions and enzymes. Plants create a large surface area in their leaves and branches, which allows them to do a great job of absorbing $CO_2$ from the atmosphere. All of the planet's trees and grasses and other biological machines pull a grand total of about 2 GT of carbon a year. To put that in context, our fossil burning is emitting 40 GT a year. Imagining that we can build machines that work 20 times better than all of biology is a fantasy created by the fossil-fuel industry so they can keep on burning.

When considering carbon sequestration, we should first remind you just how staggering that 40 GT of $CO_2$ is. If you had a giant set of scales and put all the things humans make or move on one side, and all of the $CO_2$ we produce on the other, the $CO_2$ would weigh more (see figure 17.1).

The worst version of carbon sequestration is the most seductive one: capturing $CO_2$ from thin air. This is energetically difficult, and by that I mean as difficult as juggling babies, bowling balls, electric chainsaws, and flaming tiki torches. You have to sort through a million molecules to find the 400 that are carbon, then convince those 400 to become something they don't naturally want to be: a liquid or, better yet, a solid. That sorting and conversion costs energy—a lot of it. Even if we could make it work reasonably, we'd have to install zero-carbon energy to run it, which is like using zero-carbon energy to supply our energy needs anyway, except it's more complicated and expensive to add the carbon-sequestration step. The government should fund sequestration research, within reason and with some skepticism, understanding that it's a miracle technology that we'd like to have but don't technically need, and probably can't afford.

The challenge of air capture is like a treasure hunt looking for $CO_2$ needles in the atmospheric haystack. You have to look at 2,500 molecules before you find 1 $CO_2$ molecule. For context, it is far easier to find Waldo, who in his various books appears at concentrations of around 1,200 to 4,500 PPM (or more accurately WPP, Waldos Per People).[1]

More seriously, the paper on the topic that I think is the most informative is that by Kurt Zenz House and his colleagues.[2] House analyzes carbon capture from chemistry-first principles and places a very high bar on anyone claiming to be able to sequester carbon dioxide from ambient air in a cost-effective manner. They project it would likely cost $1,000 per ton of $CO_2$; the most optimistic estimate is $300 per ton. Using the likely overly optimistic number, that would be the equivalent to adding 30¢/kWh to the cost of coal-fired electricity, or 15¢/kWh to the cost of natural gas. We should invest our time and money in things that are going to work instead.

A slightly better idea is capturing the highly concentrated $CO_2$ gas in a smokestack and somehow burying it. It is a little bit easier than the troubled idea of atmospheric $CO_2$ separations, because for some fossil fuels you can start with a concentrated flow of $CO_2$ in the smokestack, instead of a dilute gas that must be filtered from the atmosphere.

Sounds promising. But when we burn fossil fuels, we mix them with oxygen (that's what combustion is), and in so doing the burned fuels become much larger (and also a gas which makes them larger still). The idea behind carbon sequestration of fossil fuels is basically to stuff the carbon back in the hole in the ground from whence it came. But even if you squeeze carbon dioxide back down into liquid form, which costs you yet more energy and money, the volume is much larger (around five times greater) than the volume that you originally took from the ground. That's because when it came up it was mostly carbon, and when it goes back it is carbon with lots of oxygen. People propose putting carbon in other underground reservoirs, or at the bottom of the sea where the pressure of the water could contain it. But if you spring a leak, you lose all that hard work.

The economic argument against sequestration is that renewables are already competitive with coal and natural gas in most energy markets, and the added expense of carbon sequestration is not going to help fossil fuels compete. It is not unreasonable to say that the expense of carbon sequestration would be the death knell of fossil fuels.

Even though smokestack sequestration is a bad idea, the fossil-fuel industry is happy to have the American public confuse that bad idea with the worse idea of capturing the more diffuse emissions from cars, furnaces, or kitchen stoves. Those emissions are extremely distributed—they are generated at the furnace and stovetop ends of the 4.4 million miles of the US natural-gas pipeline distribution network and our 260 million tailpipes. It is nearly unimaginably difficult to collect the $CO_2$ from those sources and render it into a form that doesn't end up in the atmosphere.

In addition to the obvious business-as-usual reasons for the fossil industry to champion fossil fuels with carbon sequestration, the self-interest goes further. By injecting this $CO_2$ into the ground, the industry can force more fossil fuels back up; in fact, most of the $CO_2$ that humans have sequestered so far has been used to help with "enhanced" oil and fossil fuel recovery—further perpetuating our reliance on fossil fuel. This is an expensive, multi-layered cake of bad ideas topped with cynical frosting.

**Frack 'em all.**

## YES, AND . . . WHAT ABOUT NATURAL GAS?

Natural gas sounds benign, like the energy version of organic kale. But despite the "natural" label, it's largely methane, mixed with ethane, propane, butanes, and pentanes. When natural gas burns, like other fossil fuels, it emits carbon dioxide, carbon monoxide, and other carbon, nitrogen, and sulfurous compounds into the atmosphere, contributing to the global greenhouse-gas effect and local air pollution. Don't be fooled by those who will profit from confusion by promoting ideas like natural gas as a "bridge fuel" to the clean-energy future. Coal gets more air-time as a dirtier fuel, but natural gas is just as filthy if you account for the fugitive emissions. Natural gas is an unsafe, collapsing bridge to nowhere. We burned that bridge . . . with natural gas.

## YES, AND . . . WHAT ABOUT FRACKING?

Fracking—or hydraulic fracturing—is the process of pumping pressurized liquid into well holes to fracture the surrounding rock, which enables gas and other hydrocarbons to be more readily extracted. This technology, and the accompanying revolution of horizontal drilling, gave the US cheap natural gas at exactly the wrong moment in history.

Fracking spews methane directly from the mining sites, which offsets the nominal win from burning natural gas instead of coal. It also leaks from its network of distribution pipes. There are many other underlying problems with mining natural gas, such as water-table pollution and the creation of seismic instabilities. What's more, it's a huge distraction from the things that we know to be zero-carbon, like solar, wind, nuclear, pumped hydro, electric vehicles, and heat pumps.

## YES, AND . . . WHAT ABOUT GEOENGINEERING?

We already are geoengineering, we are just doing it badly—we're heating the earth and destroying the planet's lungs. Burning fossil fuels is geoengineering that gives us climate change. The question is, can we geoengineer for good instead?

Geoengineering is not a decarbonization strategy. It is a hope to control the temperature of the earth while giving up on $CO_2$ strategy. Many of the early arguments for studying geoengineering were that we should know how to do it, just in case the world turns out to be apathetic about climate change. We now know of multiple paths to geoengineering to mitigate climate change: most of them amount to managing the incoming flux of energy from the sun. You have probably heard of these ideas— giant space mirrors, scattering reflective particles in the atmosphere, or artificially generated clouds. In an ecosystem as complex as that of earth, all of these ideas will produce unintended effects.

Embracing geoengineering would also make us forever dependent on future geoengineering solutions. It's a bit like using liposuction as the solution to obesity when you're just going to keep eating cheeseburgers. Even if it works, we can't afford to lose sight of the better, cleaner solutions proposed in the rest of this book.

The problems of trying to control the climate are many. Who sets the temperature? Low-lying islanders and people who love coral or northern Europeans who might benefit from a bit more climate change? We don't really know all of the unintended consequences—environmental, social, or political—that would arise from geoengineering.

It is a good idea to study geoengineering schemes, and it does help us understand earth systems better, but this is not a realistic or permanent solution. It could also draw large amounts of resources away from technologies we already know can solve the problem.

## YES, AND . . . WHAT ABOUT HYDROGEN?

Many people believe hydrogen will provide a pathway to decarbonization. But hydrogen is not a source of energy. You don't discover hydrogen;

it is a battery in the form of a gaseous fuel. The fossil-fuel industry is happy to promote the hydrogen fiction because the majority of hydrogen sold today is actually a byproduct of the natural-gas industry. Only a tiny amount of gaseous hydrogen exists naturally on earth. To make and store carbon-free hydrogen, we would first have to create electricity to power a chemical process called electrolysis, which is not highly efficient. Then we'd have to capture the hydrogen gas and compress it, which consumes about 10–15% more energy. Then we'd have to decompress the gas and burn it or put it through a fuel cell. We lose more energy at every step of this process.

As a battery, hydrogen is pretty ordinary; for the one unit of electricity you put in at the beginning, you probably get less than 50% out at the other side. This is called "round-trip efficiency." To run the world off hydrogen, we'd have to produce twice the amount of electricity that we currently produce, which would itself be a monumental challenge. Remember, chemical batteries typically have 95% or so round-trip efficiency.

Germany and Japan invested heavily in hydrogen because they don't have domestic natural gas and they wanted something with the energy density of gasoline. In theory, hydrogen has about three times more energy per kilogram than gasoline (123 MJ/kg as compared to 44). But you have to compress it and store it in a tank made of exotic materials. The tank weighs much more than the hydrogen gas itself. If you include the tank in your calculations, hydrogen ends up being about a quarter of the energy density of gasoline and only a little more energy dense than batteries.

I started a company called Volute that built better compressed natural gas and hydrogen tanks. This technology is now licensed into both of those industries, so even as someone who would profit greatly from a hydrogen economy, I'm pretty confident it will only end up being a niche player. We can argue about the size of the niche. For example, hydrogen can serve as the high-temperature gas for industrial processes such as steelmaking and can solve some transportation problems.

Hydrogen will be useful, but it is not *the* answer.

## YES, AND . . . WHAT ABOUT A CARBON TAX?

A carbon tax isn't a solution. A carbon tax is a market fix meant to make all of the other solutions more competitive. It's designed to slowly increase the price of carbon dioxide, making fossil fuels uncompetitive. The idea is that a high enough carbon tax would make all of the fossil fuels more expensive than at least some of the other solutions, and then a perfectly rational market would use those cheaper clean-energy solutions.

Carbon taxes might have been sufficient if we'd started with them in the 1990s, but for the taxes to achieve the 100% adoption rates we need now, they would have to ramp up very quickly. They would also be difficult to implement, as well as regressive, hitting lower-income people hardest.

It would probably be just as effective to eliminate fossil-fuel subsidies, which in many markets would tip the scales in favor of alternatives anyway. And by the time we have the political will to implement a carbon tax, renewables with batteries will be cheaper than fossil fuels.

A carbon tax is useful in decarbonizing the hard-to-reach end points of the material and industrial economy, but unlikely to be rapid enough to transition home heating to heat pumps, and vehicles from internal combustion engines to electric vehicles at the rate required.

## YES, AND . . . WHAT ABOUT TECHNOLOGICAL MIRACLES?

"Miracle" technologies include fusion, next-generation nuclear fission, direct solar rectification, airborne wind energy, high-efficiency thermo-electric materials, ultra-high-density batteries, and other technological breakthroughs we can't yet imagine. All of these miracle technologies would, in fact, help with various components of decarbonization, and the US should invest in them as research topics. With good management, some of them might come to fruition. However, it would be unwise to bet our future on miracles, as our timeline for climate-change solutions is too short. Any ambitious technology like these would take decades to develop and scale up. We don't have decades.

The real miracles are that solar and wind are now the cheapest energy sources, electric cars are better than ICE vehicles, electric radiant heating is cozier than our existing heating systems, and the internet was a practice run and blueprint for the electricity network of the future.

## YES, AND . . . WHAT ABOUT THE EXISTING UTILITIES?

There is no way we win this war without the utilities. We need them to deliver three to four times the amount of electricity they do today. They are perfectly poised to be a giant participant in our clean-energy future.

Utilities should be the natural leaders in this project, as they already have five valuable characteristics (thanks to Hal Harvey for pointing this out): 100% market penetration, 100% billing efficacy, 100% knowledge of how we use electricity today (if they want to know it), access to low-cost capital, and an incredible local workforce in every zip code.

Beware the utility that prioritizes its natural gas business over its electricity business. If you really want to make a difference, get yourself elected to the board of your state's utility commission and steer it in the right direction.

## YES, AND . . . WHAT ABOUT EMISSIONS THAT ARE NOT ENERGY-RELATED?

This book principally concerns itself with the approximately 85% of greenhouse-gas emissions related to the US energy system.[3] They are the overwhelming majority of our emissions. The other emissions come from the agricultural sector, land use and forestry, and from industrial non-energy-use emissions. If we undertook the mobilization to address climate change as suggested in this book, this would also address much of the industrial non-energy emissions and a little of the other two, as well. Decarbonizing our energy supply is 85% of what we need to do. For the other 15%, people are successfully making and selling synthetic meats, creating pathways to cooling without terrible refrigerant emissions, and working on steel production with hydrogen and aluminum without $CO_2$. I have to believe that if we commit to the 85%, the smart and passionate people working on the other 15% will do their part, too, as I mentioned in chapter 1.

## YES, AND . . . WHAT ABOUT AGRICULTURE?

The moonshot to ignite the heartland's creativity is replacing a harmful monoculture system with an agriculture that sequesters carbon and heals our soils while also preventing the pesticide and fertilizer run-off that is polluting our rivers, estuaries, and oceans. Our world-class system of land-grant universities should be able to knock this out of the park.

## YES, AND . . . WHAT ABOUT MEAT?

There are a number of problems with meat, as any vegan will tell you. One is the amount of land required to grow the animal feed. Another is that ruminants such as cows and sheep belch methane, which is far worse as a greenhouse gas than $CO_2$. Eating less meat remains one of the easiest consumer decisions to reduce climate impact, but it alone cannot solve our climate problem. On an infrastructure scale, better land management and new low-carbon farming alternatives will lower the impact of occasional meat consumption. My old friend David MacKay used to quip that the best way to harness solar energy in Scotland was to grow and eat sheep. Meat-eating doesn't have to go away completely, but Americans do need to become more conscious about their diets.

## YES, AND . . . WHAT ABOUT ZERO-ENERGY BUILDINGS?

Building standards for extremely efficient homes that need no net-energy input, such as the energy-efficient German "passivhaus," are a good idea. Exactly what constitutes "no net-energy input" is up for debate because of the complexities of tracing material and energy flows. Some will argue that with a sufficiently good passivhaus you do not need heat-pump heating; that may be true, but we have to solve this problem for the houses that are already built as well as the houses we build tomorrow—in the US only 1% of our housing stock is built new each year.

These houses, no matter how they are built, will be rare birds. Remember, too, that only about 2% of houses are built by an architect; the majority are built from common plans by a contractor. I think of passivhaus and other similar architectural plans as a wonderful library of very good ideas for building efficient houses, and even some retrofits, and we all, especially architects and builders, should embrace these ideas and create even more.

What would perhaps have more impact in this area are the cultural shifts that make living in smaller, simpler houses more desirable. Mobile homes have gotten a bad cultural rap, but they have a smaller carbon footprint than conventional houses and could offer one of the fastest pathways for adopting modern decarbonized domestic infrastructure.

## YES, AND . . . WHAT ABOUT THE REST OF THE WORLD?

America is only responsible for about 20% of current annual global emissions (though historically, it has produced a larger share). People say this is why it's not worth bothering with decarbonizing the US. China will emit more, and if not them, it will be the Saudis, or India, or Africa. If we all adopt that defeatist attitude, we are done. If America leads, however, it is likely that other countries will follow once they see the economic advantages of doing so. The early movers will own the lion's share of these critical twenty-first-century industries.

## YES, AND . . . CAN WE MAKE ENOUGH BATTERIES?

No two ways about it, we will need a lot of batteries. This is not impossible, though, given America's current manufacturing capacity. To replace our 250 million personal gasoline-powered vehicles with EVs in the next 20 years, we will need over a trillion batteries, or around 60 billion 18650 batteries every year (18650's are 18 millimeters in diameter and 65 millimeters long—slightly larger than your flashlight's AA batteries). That is similar to the 90 billion bullets manufactured globally today. We need batteries, not bullets.

## YES, AND . . . WHAT ABOUT FLYING?

Flying is energy-intensive per minute, but not per mile. Per passenger-mile traveled, it requires approximately the same energy as driving in a car with a passenger. That said, reducing the number of flights taken is one of the most effective ways for individuals to reduce their energy footprints.

In the electrified future, short-haul flights (under 500 miles) will be electric, enabled by increases in the power density of motors and batteries. Long-haul flights will use biofuels to get enough range. Passenger and freight flights in the US require a total of 2 quads, and military aviation is another 0.5 quads. The US can produce about 10 quads of biofuel energy, easily covering the tab for flying, in addition to other hard-to-electrify things like construction and mining equipment (which together add another 1–2 quads).

I have several friends who have electric aircraft companies; they are very bullish on flying cars. I have another colleague who accurately states that at about 80 mph it starts to take more energy to keep the car on the ground than just flying it—keeping the car's tires on the ground costs you a lot in energy! It is even possible to convince yourself that small electric aircraft will have energy efficiencies per passenger mile similar to electric cars. This is true if you fly naked, but not if you pack a lot of luggage. Also, if we could all fly everywhere quickly, we'd do it more, and lose the gains in extra miles traveled. As a result, I predict this will remain the domain of billionaires.

## YES, AND . . . WHAT ABOUT AUTONOMOUS CARS?

Like flying cars, autonomous cars have captured the public's imagination (not to mention the self-interested parties trying to profit from them). Supposedly, they will reduce traffic and lower emissions. This is almost certainly not true. When groups of people were given a chauffeur as a stand-in for an autonomous vehicle, they drove many more trips, and would occasionally send the "autonomous" car across town to buy them their favorite sandwich.[4] Autonomous cars will almost certainly induce more miles traveled.

In the taxi industry, there is something known as "carriage-miles." This is the ratio of miles driven without a passenger to miles driven with one. For taxis, this ratio is about 1.7, meaning the car has to drive 1.7 miles to move a passenger 1 mile. In disrupting the taxi industry, Uber and Lyft were able to get this number down to about 1.4. This is probably a good proxy for what will happen with wide deployment of autonomous vehicles. Even if we all are driven around to the same places, we'll increase miles driven by 40%. Honestly, this is yet more Silicon Valley snake oil.

## YES, AND . . . WHAT ABOUT THE DANGERS OF NUCLEAR POWER?

America has led the world in nuclear power. The US Navy operates the largest fleet of small reactors in the world, and it boasts an impeccable safety record. Nuclear is a form of electrification, and it fits squarely with the plan to fight global heating. Nuclear power currently delivers around 100 GW of very reliable electricity to America's grid. Maintaining or even ambitiously increasing this amount would no doubt make the climate solution easier. Today's best estimates have nuclear energy at approximately double the cost of wind and solar. Without a doubt, those costs could be trimmed enormously given advances in engineering, since most of these plants were designed 50 years ago.

The health effects of nuclear power have been well studied. It is established that nuclear is not as dangerous as we tend to think. But like shark attacks, it's the prospect of a low-probability event that could release radiation that drives our fears. We can lower that probability further by building dedicated infrastructure like the facility at Yucca Mountain, but the fact remains that for 40 years, policymakers haven't been sufficiently able to convince people to invest in this kind of infrastructure. Nuclear power will remain a very difficult political topic unless we have a breakthrough in waste management.

## YES, AND . . . WHAT ABOUT GROWING TREES?

Yes, we should—at least a trillion. Grab a shovel!

The best time to plant a tree is 30 years ago. The second-best time to plant a tree is today.

Go plant a tree for your grand-kids to climb on. Even better, go plant 30,000.

# B

## WHAT CAN YOU DO TO MAKE A DIFFERENCE?

It's time to get to work. Ask not what your planet can do for you but what you can do for your planet. Everyone has a role.

Your first role is as a citizen. Become a political agitator, work on things that make a difference, embrace twenty-first-century solutions to twenty-first-century challenges. Lots of things will change; only be nostalgic for the things that truly matter. To fix climate change, we need unlikely coalitions. We need to bring people to the table from all walks of life—urban and rural, government and business, red state and blue state, black and white, union members and gig workers, young and old.

If you are eligible to vote, it is time to vote for politicians who take climate change seriously. If you do support these politicians, and they enact a plan as ambitious as the one outlined in this book, there is a glorious future for us all. If you do not, the next hundred years are going to be pretty grim. As the COVID-19 pandemic reminded us, threats that seem remote and distant can erupt far more suddenly than we expect. Just as the prospect of a pandemic seemed like something that *could* happen, but somehow seemed unlikely despite experts' warnings, a far larger storm is brewing, and it's well past time to begin preparing.

An event like the pandemic may hurt the old economy, but it is a transformative opportunity for a new one. We can maintain this economy that lurches from one predicted but unplanned disaster to the next, only to find ourselves mid-century in an endless string of climate change–induced disasters that frankly make COVID look like a picnic. Or we can wake up now and get to work building a better future. This project has the capacity to be the foundation of a new American economy that employs more people, and in better jobs, than ever before.

**If you are not old enough to vote**, you should vote with your feet by protesting. The youth climate strike is a fabulous place to start. You might also consider various ways to file lawsuits against the adults and industries that are stealing your future. Get angry and get creative, but remember to have fun and forge great friendships along the way. Chain yourself to a fence. Fall in love with the passionate activist beside you.

**If you are a consumer**, don't focus so much on your small decisions. While it may be helpful to buy shampoo in bulk to eliminate the plastic or buy all-natural clothes that can be composted, what matters most are your big purchasing decisions. Your next car must be electric. You need to do everything you can to make your house run on solar power. If you are about to buy a house, consider a smaller one or a mobile home. Whatever you invest in turning your house into a big battery that can give back to the grid will have more impact on climate change than any other purchasing decision you make.

**If you are a farmer**, this is an incredible opportunity to reimagine agriculture. American farmers and their incredibly productive lands are central to global climate success. Let's make our lands generative and let's make them absorb carbon in the soils, not release it.

**If you are an engineer**, there is a lot of work to do. Get to work hammering out the details of our electrified future. Design the new grid. Make things more reliable, robust, and affordable. Squeeze out the last few percentage points of performance.

**If you are a lawyer**, you should either file a lawsuit against the fossil-fuel interests or you should work to overturn local ordinances and building codes that are impediments to rolling out climate solutions as quickly and cheaply as possible.

**If you are a small-business owner**, differentiate your product by making it cleaner and greener sooner. Make products that every American wants.

**If you run a school or community college**, we need more shop classes and more trainees in the practical arts. We need more Americans who know how to make and install things, turn screws and tighten bolts, and build the future.

**If you are a designer**, make electric appliances so beautiful and intuitive that no one would ever buy anything else. Design electric vehicles that redefine transport. Create products that don't need packaging. Make products that want to be heirlooms.

**If you are a union representative**, don't let a fear of lost jobs stand in the way of the enormous number of jobs that will be created in a zero-carbon economy. Prepare yourself and your union by working with environmental lobbyists for guaranteed job placements, transfer of pay and benefit levels, and retraining programs. Without labor, there will be no economic transformation.

**If you are a teacher or professor**, communicate clearly to your students the intergenerational burden that has been placed upon them. Teach them about science and justice and inspire them all to be activists. Most of all, you need to help your charges understand that no one is coming to save us; we must save ourselves.

**If you are a poet or an artist**, we desperately need love letters to planet earth. Inspire us with beauty to appreciate the world and each other. Help us ask the right questions.

**If you are an investor**, invest in companies that are working toward a carbon-free future. Divest from fossil fuels. Be less greedy. Remember that profits mean nothing if the planet is ruined.

**If you are an electrician**, prepare to be the busiest you have ever been. Train your friends, teach your children.

**If you are a roofer**, learn to be a solar installer, too, and prepare for a big increase in demand.

**If you are an hourly worker**, advocate for the renewable economy, because your wages will go up. Better jobs are coming if we get this right.

**If you are in construction or renovation**, encourage your clients to shift to houses that don't pipe in natural gas and buildings that are solar-powered. Learn to install heat pumps and batteries that make houses run efficiently.

**If you are an architect**, this is a great time to work on propagating new architectural solutions that maximize a building's potential to be part of the solution. This means rooftops that are flatter and face toward the sun (south-facing in the Northern Hemisphere). It means promoting high-efficiency houses, lighter construction methods, and, given that buildings use so many materials, finding ways for the buildings to be net absorbers of $CO_2$ rather than net emitters.

**If you are an entrepreneur**, start the billion-dollar clean energy company that addresses 0.5% of our energy economy. We only need 200 of you to succeed.

**If you are a doctor or health care professional**, speak loudly and clearly about the human costs of pollution and fossil fuels. Respiratory illness caused by burning fossil fuels kills millions of people globally. Asthma, bronchitis, and pneumonia are exacerbated by the particulate matter created by burning dead dinosaurs. Cancers proliferate that were caused by hydrocarbons, dioxins, and other chemicals born of the fossil fuel economy. Sedentary, car-based lifestyles lead to obesity, diabetes, heart disease, and other serious ailments. Vastly better public health outcomes will be gained by transitioning rapidly to a clean-energy world.

**If you are a mechanic**, start building electric hot-rods. After all, it's the sheet metal that we fall in love with, not the engine.

**If you are a biologist**, help make biofuels and biomaterials to power long-distance flights and sectors of the economy that can't run on wind, solar, or nuclear energy.

**If you are a tech worker**, stop making social media and delivery apps and start making software that helps people use less energy and that balances the grid, automates the design of solar and wind plants, makes public transit work better, and does other useful things to accelerate America's transition to renewables.

**If you are a social worker**, you can be an advocate for helping lower-income people access homes and transportation using clean-energy sources.

**If you are a city planner**, help make US cities and towns more amenable to a zero-carbon future.

**If you are a coal miner**, thank you for your service. Now you'll have a job mining materials for batteries and electric motors.

**If you are an oil-industry worker**, thank you, too, for your service. Now you'll have a job helping America build the massive infrastructure that is required for a zero-carbon future.

**If you are a politician**, you need to listen to people in this order: scientists, children, engineers. Then you need to rise above the din and clear the regulatory and financial paths required to get this job done. Work with everyone. Redefine political boundaries, parties, and coalitions.

**If you are a city, town, or county representative**, listen to your constituents and find out what is holding them back from buying electric vehicles, installing solar power, purchasing clean energy from their utilities, retrofitting their houses, and securing loans to buy decarbonization technologies for their home. Remove all of these barriers by whatever means necessary.

**If you are a mayor**, change the local building codes as necessary to promote the fastest, cheapest ways to decarbonize. Install clean energy on local buildings. Create electric-vehicle infrastructure everywhere in your town.

**If you are a state-level politician**, it is useful to remember that the states are experiments. No one has the perfect answer to decarbonization, and we all have things to learn from one another. Be bold, take risks, and write the brilliant piece of legislation that speeds up the clean-energy transition that can be cut and pasted into other state's policies and even federal programs.

**If you are a congressperson or senator**, stand up to corrupting influences. Remember that you were elected by people, not corporations, and that you were elected to improve Americans' lives for the long term.

**If you are a president**, lead. With vision. Try some FDR, a dollop of Churchill, a dash of JFK, a pinch of Reagan, a seasoning of Mandela, and a splash of Merkel.

**If you are a corporate CEO**, you should be leading your company with an authentic vision for the future and preparing to fully decarbonize your operations in a decade. You will need to listen to your youngest employees, as well as the frustrated older ones who have been telling you to change for years. Between those groups, you probably already have the solutions within your organization. Stop worshipping the quarterly numbers and build your company for the future.

**If you are a billionaire**, you might consider buying out a fossil-fuel lease or two. Own a piece of history in some remote place. Turn it into a nature reserve. Divest your portfolio of fossils. Invest in start-ups that offer ambitious solutions, even if they can't offer fast, guaranteed returns. Sponsor young activists. Lose some money swinging for the fences as though you were 24 years old again. You have nothing to lose other than the planet.

**If you are a vegan cyclist**, thank you. Live long and prosper.

**If you are a singer or songwriter**, nothing is more powerful than music to move people. We need anthems for our movement. We need some Neil Young—he had his vintage Lincoln Continental electrified to demonstrate his commitment to the future. Throw in some Cat Stevens and some Joni Mitchell, too.

Don't it always seem to go
That you don't know what you've got 'til it's gone
They paved paradise
And put up a parking lot[1]

**In building an abundant and verdant future, there is a job for everyone. Good luck. May the winds be with us.**

# C

# DOWN THE RABBIT HOLE: CLIMATE SCIENCE 101

---

☞ Climate science encompasses a broad array of activities and a lot of layers of varying complexity.

☞ Basic climate science seeks to understand the underlying physics and chemistry of earth systems with careful measurements.

☞ Climate modeling seeks to assemble the findings of basic climate science into models of how earth systems interact.

☞ The first climate model was built without computers and could predict climate change reasonably well.

☞ Impact studies predict the social, economic, political, and other impacts based on the climate model.

☞ Carbon budgets are the estimated further emissions that we can afford given a particular temperature or climate target.

☞ Emissions trajectories are the projections for emissions reductions and technology changeover required to hit a carbon budget.

☞ Integrated assessments try to put all of the pieces together in reports for broader audiences.

☞ The science is sound, and we have the tools that tell us what we need to do.

Getting from climate science to climate action requires several steps, which I will walk through here.

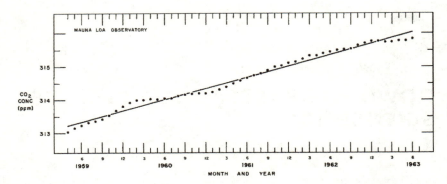

**20.1** The original "Keeling Curve" of atmospheric $CO_2$ concentration, and its continuation. Source: Jack Pales and Charles Keeling, "The Concentration of Atmospheric Carbon Dioxide in Hawaii," *Journal of Geophysical Research* 70, no. 24, 1965.

## CLIMATE SCIENCE

First, we must do climate science, the detailed work of measuring and modeling the clouds, glaciers, oceans, soils, emissions, and other factors affecting global climate. These component systems are simple enough that the physics can be deconstructed and probed, generating predictions that can be confirmed through measurements.

For example, in one of the most iconic climate science studies, Pales and Keeling first documented the increasing concentration of $CO_2$ in the atmosphere.[1] The resulting "Keeling Curve" of $CO_2$ concentration, as it is now known, is shown in figure 20.1. Through fastidious measurements taken between 1959 and 1963 atop an observatory at Hawaii's Mauna Loa volcano, the study showed both seasonal absorption of $CO_2$ by trees, as well as the disturbing and long-term upward trend due to burning fossil fuels. The measurements of that study have continued to be collected since then, providing further documentation of these trends.[2]

## CLIMATE MODELING

After climate science, we must perform climate modeling, the assembly of these components into a model of the entire climate system. The interactions of models are complex, and system-level models today typically employ large computers to crunch the numbers. These models are quite

TABLE VII.—*Variation of Temperature caused by a given Variation of Carbonic Acid.*

| Latitude | Carbonic Acid = 0·67. | | | | | Carbonic Acid = 1·5. | | | | | Carbonic Acid = 2·0. | | | | | Carbonic Acid = 2·5. | | | | | Carbonic Acid = 3·0. | | | | |
|---|---|---|---|---|---|---|---|---|---|---|---|---|---|---|---|---|---|---|---|---|---|---|---|---|---|
| | Dec.-Feb. | March-May. | June-Aug. | Sept.-Nov. | Mean of the year. | Dec.-Feb. | March-May. | June-Aug. | Sept.-Nov. | Mean of the year. | Dec.-Feb. | March-May. | June-Aug. | Sept.-Nov. | Mean of the year. | Dec.-Feb. | March-May. | June-Aug. | Sept.-Nov. | Mean of the year. | Dec.-Feb. | March-May. | June-Aug. | Sept.-Nov. | Mean of the year. |
| 70 | -2·9 | -3·0 | -3·4 | -3·1 | -3·1 | 3·3 | 3·4 | 3·8 | 3·6 | 3·52 | 6·0 | 6·1 | 6·0 | 6·1 | 6·05 | 7·9 | 8·0 | 7·9 | 8·0 | 7·95 | 9·1 | 9·3 | 9·4 | 9·4 | 9·3 |
| 60 | -3·0 | -3·2 | -3·4 | -3·3 | -3·22 | 3·4 | 3·7 | 3·6 | 3·8 | 3·62 | 6·1 | 6·1 | 5·8 | 6·1 | 6·02 | 8·0 | 8·0 | 7·6 | 7·9 | 7·87 | 9·3 | 9·5 | 8·9 | 9·5 | 9·3 |
| 50 | -3·2 | -3·3 | -3·3 | -3·4 | -3·3 | 3·7 | 3·8 | 3·4 | 3·7 | 3·65 | 6·1 | 6·1 | 5·5 | 6·0 | 5·92 | 8·0 | 7·9 | 7·0 | 7·9 | 7·7 | 9·5 | 9·4 | 8·6 | 9·2 | 9·17 |
| 40 | -3·4 | -3·4 | -3·2 | -3·3 | -3·32 | 3·7 | 3·6 | 3·3 | 3·5 | 3·52 | 6·0 | 5·8 | 5·4 | 5·6 | 5·7 | 7·9 | 7·6 | 6·9 | 7·3 | 7·42 | 9·3 | 9·0 | 8·2 | 8·8 | 8·82 |
| 30 | -3·3 | -3·2 | -3·1 | -3·1 | -3·17 | 3·5 | 3·3 | 3·2 | 3·5 | 3·47 | 5·6 | 5·4 | 5·0 | 5·2 | 5·3 | 7·2 | 7·0 | 6·6 | 6·7 | 6·87 | 8·7 | 8·3 | 7·5 | 7·9 | 8·1 |
| 20 | -3·1 | -3·1 | -3·0 | -3·1 | -3·07 | 3·5 | 3·2 | 3·1 | 3·2 | 3·25 | 5·2 | 5·0 | 4·9 | 5·0 | 5·02 | 6·7 | 6·6 | 6·3 | 6·6 | 6·52 | 7·9 | 7·5 | 7·2 | 7·5 | 7·52 |
| 10 | -3·1 | -3·0 | -3·0 | -3·0 | -3·02 | 3·2 | 3·2 | 3·1 | 3·1 | 3·15 | 5·0 | 5·0 | 4·9 | 4·9 | 4·95 | 6·6 | 6·4 | 6·3 | 6·4 | 6·42 | 7·4 | 7·3 | 7·2 | 7·3 | 7·3 |
| 0 | -8·0 | -3·0 | -3·1 | -3·0 | -3·02 | 3·1 | 3·1 | 3·2 | 3·2 | 3·15 | 4·9 | 4·9 | 5·0 | 5·0 | 4·92 | 6·1 | 6·1 | 6·6 | 6·6 | 6·5 | 7·3 | 7·3 | 7·4 | 7·4 | 7·35 |
| -10 | -3·1 | -3·1 | -3·2 | -3·1 | -3·12 | 3·2 | 3·2 | 3·2 | 3·2 | 3·2 | 5·0 | 5·0 | 5·2 | 5·1 | 5·07 | 6·6 | 6·6 | 6·7 | 6·7 | 6·65 | 7·4 | 7·5 | 8·0 | 7·6 | 7·62 |
| -20 | -3·1 | -3·2 | -3·3 | -3·2 | -3·2 | 3·2 | 3·2 | 3·4 | 3·3 | 3·27 | 5·2 | 5·3 | 5·5 | 5·4 | 5·33 | 6·7 | 6·8 | 7·0 | 7·0 | 6·87 | 7·9 | 8·1 | 8·6 | 8·3 | 8·22 |
| -30 | -3·3 | -3·3 | -3·4 | -3·4 | -3·35 | 3·4 | 3·5 | 3·7 | 3·5 | 3·52 | 5·5 | 5·6 | 5·8 | 5·6 | 5·62 | 7·0 | 7·2 | 7·7 | 7·4 | 7·32 | 8·6 | 8·7 | 9·1 | 8·8 | 8·8 |
| -40 | -3·4 | -3·4 | -3·3 | -3·4 | -3·37 | 3·6 | 3·7 | 3·8 | 3·7 | 3·7 | 5·8 | 6·0 | 6·0 | 6·0 | 5·95 | 7·7 | 7·9 | 7·9 | 7·9 | 7·85 | 9·1 | 9·2 | 9·4 | 9·3 | 9·25 |
| -50 | -3·2 | -3·3 | — | — | — | 3·8 | 3·7 | — | — | — | 6·0 | 6·1 | — | — | — | 7·9 | 8·0 | — | — | — | 9·4 | 9·5 | — | — | — |
| -60 | | | | | | | | | | | | | | | | | | | | | | | | | |

**20.2** Arrhenius's 1897 model of temperature variation versus concentration of carbonic acid ($CO_2$). Source: Svante Arrhenius, "On the Influence of Carbonic Acid in the Air upon the Temperature of the Ground," *Astronomical Society of the Pacific* 9, no. 54, 1897.

good now, and have been tested rigorously against past data to verify predictive accuracy. There is still some uncertainty, but a small and quantified amount, and the major phenomena are well agreed upon.

For example, in one of the first papers to model the climate, Swedish Nobel Prize–winning scientist Svante Arrhenius showed the relationship between increasing $CO_2$ concentration and temperature (shown in figure 20.2). In the 120 years since then, climate scientists have been expanding on this simple model to increase its scope by leveraging the ever-growing computational resources available (for instance, the seminal 1967 paper by Manabe and Wetherald that established temperature equilibrium conditions for the atmosphere[3]). So while the best models today embody great complexity, the punchline was simple enough to be worked out on paper: increasing $CO_2$ concentration leads to increasing temperature.

## IMPACT STUDIES

After modeling the climate, scientists undertake impact studies to determine how the climate affects other things we care about, like humans, geography, animals, earth systems, economics, and pandemics. Impact studies warn us what will happen as a result of climate change. They tell us how much sea level change will occur for a given emissions trajectory, and how many people will be displaced. They tell us how climate change will affect our agriculture and food supply. They help us understand the changing patterns and intensities of events like storms and bushfires. Impact studies encompass a broad range of disciplines (including economics, politics, and engineering), and address an enormous number of issues including food security,[4] tourism,[5] poverty,[6] migration,[7] economics,[8] war,[9] air quality,[10] disease,[11] and labor,[12] among others. The breadth of these impact studies can be overwhelming, but the IPCC regularly publishes summaries for mere mortals.[13]

## CARBON BUDGETS

With our climate models and impact studies of how climate affects the things we care about, we can then draft carbon budgets, or estimates of allowable carbon emissions that limit adverse impacts to manageable levels. This shows us, in concrete terms, the amount of $CO_2$ or other greenhouse gasses we can emit.

Perhaps the most iconic carbon-budget study was the "trillion ton" paper, which gave a clear and sobering method for budgeting our remaining carbon emissions in terms of the resulting temperature rise.[14] The "trillion" study highlights that we need to stop short of 1 trillion tons of cumulative emissions.

## EMISSIONS TRAJECTORIES

With a carbon budget, we can then create an emissions trajectory, or a yearly progression of allowable emissions that hit the budget. This is not quite climate science—it's more like climate socio-economics, as it tries to estimate how human behavior will respond to fighting a warming climate. Figure 20.3 shows a graph of emissions trajectories starting at present and proceeding through 2100. We clearly see how current policies and global pledges and commitments don't come close to our

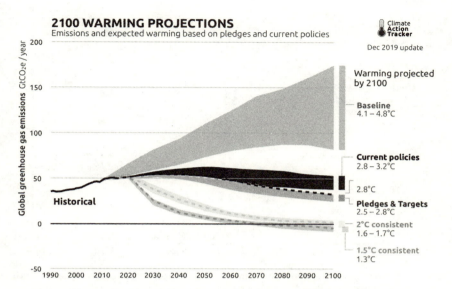

**20.3** 2100 warming projections mapped to commitments, trends, pledges, and targets. Source: Climate Action Tracker, n.d., https://climateactiontracker.org/global/temperatures/.

desired 2°C/3.6°F target let alone the 1.5°C/2.7°F target we should be aiming for.

## INTEGRATED ASSESSMENTS

Finally, people who work in climate science and policy write integrated assessments, which package all the steps into digestible reports and policy recommendations. These reports take years to assemble and include the work of hundreds or even thousands of contributing scientists. This is the work of the IPCC.

## WHERE DOES THAT LEAVE US?

There is a lot of room for confusion and opportunity for people to disagree on urgency, because this is not a simple question.

I imagine the confusion plays out like this in people's minds: They read a newspaper article about an impact study that captures their eye. This

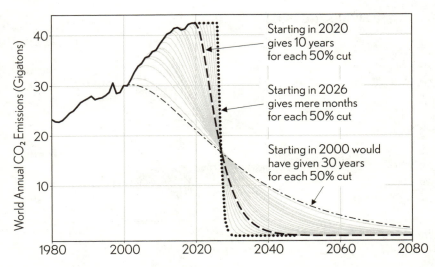

**20.4** Mitigation curves redrawn from Robbie Andrew, "It's Getting Harder and Harder to Limit Ourselves to 2°C," Desdemona Despair, April 23, 2020, https://desdemonadespair.net/2019/08/its-getting-harder-and-harder-to-limit-global-warming-to-2c-it-is-partly-this-hope-in-future-technologies-that-delays-action.html.

press article is likely a poor summary of the science behind the impact study. Maybe they go on to read the original study. Then they decide if it impacts them or something they care about—which was probably the headline that made them read the article in the first place. Some people might then try to find out how avoiding that impact relates to an emissions trajectory. By this point they're likely to get lost in the complexity of it all. If not, they likely will be at the next stage when they look into the black-hole of uncertainties related to any given emissions trajectory within a climate model.

For me, it is pretty simple. I want to live in a world that has coral reefs and rainforests full of the beautiful things I experienced as a child. I also fear the panicked responses of human beings when we feel impacts and stressors on our nations' food systems. Even a warming of 1.5°C/2.7°F will create a great deal of disruption and stress in the world, so I want to try to figure out how to keep us as close to that target as possible.

## MITIGATION CURVES

Mitigation curves are the near-term emission responses required to keep us on track for a particular climate target, such as 1.5°C/2.7°F or 2°C/3.6°F. The impact of even so much as a four-year delay on our chances is terrible, as is illustrated clearly in figure 20.4.

The simple conclusion is that we now must reduce emissions as fast as humanly possible, with an emergency, wartime-level response. We need to turn aggressive mitigation curves into an action plan with production timetables and deliverables in order not to experience the worst impacts of climate change.

# D

# DOWN THE RABBIT HOLE: HOW TO READ A SANKEY FLOW DIAGRAM

I use a lot of charts in this book, despite the fact that I'm told it is a recipe for publishing failure. That's why I'm hiding so many of them here at the back of the book. The one type of chart I refer to extensively is called a "Sankey diagram," so this is an introduction to where they came from and how to read them. They're elegant tools that can communicate complex issues—the big picture and the little details at once—and they are fairly simple to understand.

The first Sankey-style flow diagram that we know of was drawn by a French engineer named Charles Joseph Minard. In 1845, he created a flow chart to depict the traffic on roads between Dijon and Mulhouse, France, to inform the routing of a new railroad.[1] In 1869, he created the chart that he is best known for: a visualization of Napoleon's foray into Russia, his retreat, and troop losses throughout the campaign (figure 21.1).

The two-dimensional band chart displays six types of data: the number of Napoleon's troops, the distance they traveled, temperature, latitude and longitude, direction of travel, and their location on specific dates.

The pink line that starts out thick on the left is proportional to the number of troops that were alive at any point. The thinner the line, the more tragic the story. By the time he arrived in Moscow, he'd lost two-thirds of his troops, and then he was beaten all the way back, losing more men along the way, as represented by the black line. The chart shows that

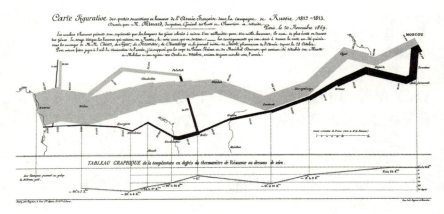

**21.1**  The often-cited 1869 "Sankey" diagram of Napoleon's troop levels as he fought his way into, and back from, Russia. Source: Sandra Rendgen, *The Minard System: The Complete Statistical Graphics of Charles-Joseph Minard* (Princeton, NJ: Princeton Architectural Press, 2018); from the collection of the École Nationale.

by the time he returned to Kowno, his 422,000 troops had been reduced to just 10,000.

A few years after Minard created his flow chart of Napoleon's 1812–1813 Russian campaign, Irish Captain Matthew Henry Phineas Riall Sankey created a flow chart to visualize the efficiency of the steam engine (that presumably powered his ship). This is the first known use of the (as it would become known) Sankey diagram to visualize the flow of energy. The width of the arrows is proportional to the flow rate of energy.

At the time, coal had become an important fuel in powering ocean-going vessels, and for very good reason. Wind doesn't always blow, and only rarely from the direction you want. Coal sits in the bottom of your hull, ready to shovel at a moment's notice. While many sailors were still mainly concerned with the weather and sails, Sankey was diagramming the conversion of coal energy into pressurized water and steam, and understanding the energy losses along the way.

Sankey diagrams are particularly good for visualizing energy because they conserve proportion at every point along the flow. This gels well with the law of conservation of energy and the related first law of thermodynamics, which both state that energy can neither be created nor destroyed, but can only change from one form to another.

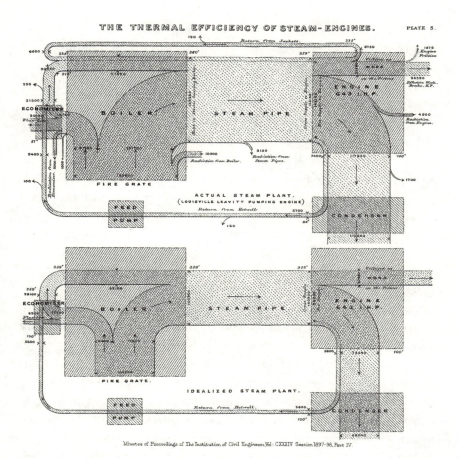

**21.2** The original Sankey diagram, by Captain Sankey. We can see from Sankey's diagram the conversion of coal energy into pressurized water and steam in the boiler, and the consequent losses as it is turned into propulsive power at the propeller of his boat. Source: Alex B. W. Kennedy and H. Riall Sankey, "The Thermal Efficiency of Steam Engines," *Minutes of the Proceedings of the Institution of Civil Engineers* 134, part IV (1898): 278–312, doi:10.1680/imotp.1898.19100.

If you look where energy is lost in the flows, it is typically to heat. This is as true today as it was for Captain Sankey; the cold hard destiny of all energy is that eventually it becomes low-grade heat from which it is too difficult to extract useful work. The temperature of the universe is 2.73°K, −270°C or −455°F—chilly no matter what unit we express it in. The fate of all of our waste heat here on earth will ultimately be to radiate its way out into space as it cools to the temperature of the universe.

This book is largely about energy budgets. To illustrate them with an analogy, most people have some understanding of their household budget, so in figure 21.3, I present the average US household budget as a Sankey flow diagram.[2]

The chart is read from left to right. The inputs to the house are things like income and interest on accounts. These flow into a total household budget. The total household budget then flows into four large categories of things: housing, transportation, food, and the catch-all bucket "everything else." These further break down into the minutiae of how we spend money—on gasoline, eating out, clothes, and other activities of our daily lives. This same data is presented in table form in figure 21.4.

The average US household is known as the "consumer unit." Consumer units include families, single persons living alone or sharing a household with others who are financially independent, or two or more persons living together who share major expenses. In 2019, the average American consumer unit had a pre-tax income of $78,635.

The average household expenditures totaled $61,224. We can see that it is spent on four large categories of things: transportation, housing, food, and everything else. Housing is the biggest category. $3,477 is spent in the home on utilities, fuels, and public services; we can already see the link between energy and our own personal budget. Transportation is another big segment; of that sector, a third flows into the cost of gasoline. The average household spends a little more than that on health care, and less on savings and retirement. We spend less again, under $1,000, on education. We spend only about $120 a year on reading.

Just as some of you might be interested in chasing down every dollar in your household budget, I'm interested in tracking down every joule of energy in the US and global economy. Sankey diagrams have been

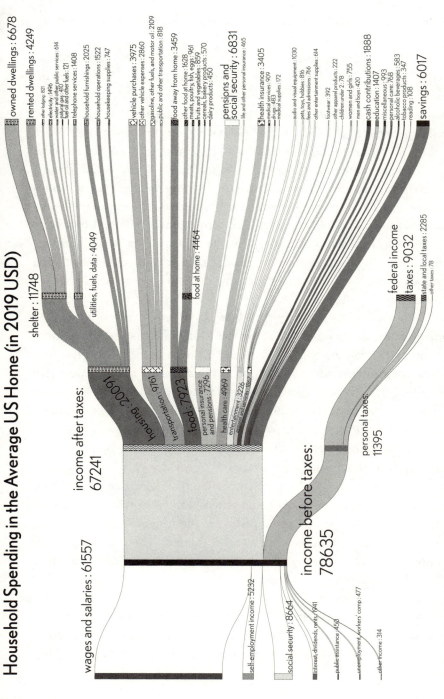

# Household Spending in the Average US Home (in 2019 USD)

**21.3** This Sankey diagram is more relatable to your everyday life. This takes Census Bureau data for household spending in the average US home and represents it as flow. Data from the US Bureau of Labor Statistics, *Consumer Expenditures Report 2019*, December 2020, https://www.bls.gov/opub/reports/consumer-expenditures/2019/home.htm.

instrumental in this analysis. Let's go for an eye-opening ride (if you can keep your eyes open!).

Sankey diagrams found widespread use during the oil crises of the early 1970s. In 1973, Jack Bridges, working for the Joint Committee on Atomic Energy, reprised the Sankey and improved upon it in a fabulous book, *Understanding the National Energy Dilemma*. The United States had just experienced oil shortages, and energy was on people's minds.

The book was quite novel. It not only employed a Sankey to show current energy trends, it provided historic Sankeys and future projections to communicate the challenges of planning and delivering a country's energy supply. A Sankey of 1950, 1960, 1970, 1980, and 1990 were included in special full-color, fold-out centerfolds. These fold-outs were intended to be assembled into a three-dimensional view of the history and future of US energy consumption. It showed the rapidly increasing total demand, divided into portions of "lost" energy and "used" energy. Lost energy is waste heat.

Place yourself in the historical context of this diagram, when the energy world was in upheaval. There was a major oil crisis because America's appetite for oil had outpaced its production, and for the first time our energy destiny, growth capacity, and entire future as a nation were tied to geopolitics. Everything was being disrupted; our desire for electricity was proving as insatiable as our desire for gasoline, and we were installing hydroelectric plants everywhere we could. Nuclear-powered electricity was just starting to take off, and estimates of its future potential were hyperbolic; it was already controversial.

Nuclear advocates at the time would say that "electricity will be too cheap to be metered." People had a renewed interest in wind power for generating electricity, and some people at the fringes were just starting to talk about solar architecture and solar thermal. The changing energy landscape, and the urgency of the oil crises, underscored the importance of tools for visualizing and planning the future of the American and global energy supply.

The methodologies behind these visualizations are now used in the EIA Annual Energy Review (AER) and in yearly summaries of the energy economy made public by Lawrence Livermore National Laboratory (LLNL).[3]

**Table 21.1**  Average Income and Expenditures of All Consumer Units, 2018

| ITEM | 2018 Expenditure |
| --- | --- |
| Average income before taxes | $78,635 |
| Average annual expenditures | 61,224 |
| Food | 7,923 |
| Food at home | 4,464 |
| Food away from home | 3,459 |
| Housing | 20,091 |
| Shelter | 11,747 |
| Owned dwellings | 6,678 |
| Rented dwellings | 4,249 |
| Apparel and services | 1,866 |
| Transportation | 9,761 |
| Vehicle purchases | 3,975 |
| Gasoline, other fuels, and motor oil | 2,109 |
| Healthcare | 4,968 |
| Health insurance | 3,405 |
| Entertainment | 3,226 |
| Personal care products and services | 768 |
| Education | 1,407 |
| Cash contributions | 1,888 |
| Personal insurance and pensions | 7,296 |
| Pensions and Social Security | 6,831 |
| All other expenditures | 2,030 |

Source: the Bureau of Labor Statistics (BLS), "Consumer Expenditures—2019,"
news release, September 9, 2020, https://www.bls.gov/news.release/cesan.nr0.htm.

# Global Exergy Accumulation, Flow, and Destruction

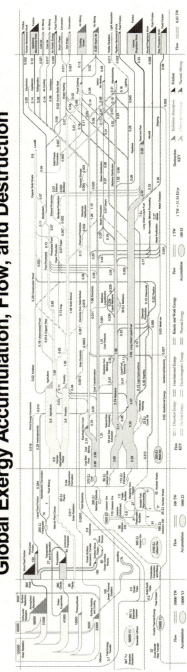

# The Natural and Anthropogenic Carbon Cycle

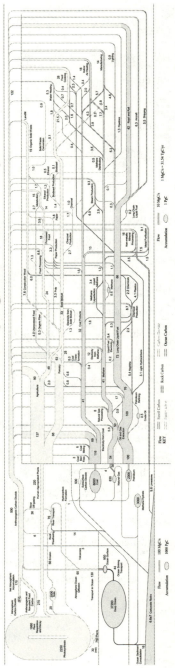

Perhaps the most intense Sankey diagram is the one by Wes Hermann, shown in figure 21.4. I was first introduced to this chart in around 2007 by Hermann, who interviewed with me for my company Makani Power. Wes didn't take the job with us, though the interview was great; he took a job with a young electric vehicle company—Tesla—instead. When I asked him about the choice, emphasizing how important wind energy was, he simply replied that burning gasoline to run cars was the dumbest destruction of otherwise useful energy in the whole system and that electric cars were the only way to go. He was right about electric cars, though wind is still key. He made the chart as a student of the Global Climate and Energy Program at Stanford. While the chart itself is nearly impossible to read

---

**21.4**  Wes Herman's Sankey charts for Exergy and Carbon. (Top) Exergy is the useful portion of energy that allows us to do work and perform energy services. While energy is conserved, its exergy content can be destroyed when the energy undergoes a conversion. We gather exergy from distinct, energy-carrying substances in the natural world we call resources. These resources are converted into forms of energy called carriers that are convenient to use in our factories, vehicles, and buildings. This diagram traces the flow of exergy through the biosphere and the human energy system, illustrating its accumulations, interconnections, conversions, and eventual natural or anthropogenic destruction. Choices among exergy resources and exergy carriers and the manner of their utilization have environmental consequences. An examination of the available resources and current human exergy use places the full range of energy options available to our growing world population and economy in context. This perspective may assist in efforts to both reduce exergy use and decouple it from environmental damage. (Bottom) The ability of carbon-based molecules to store a large amount of useful energy, or exergy, has made carbon an important exergy carrier in natural and human energy systems. Carbon is being reintroduced to the biosphere through the human use of fossil fuels at a rate far exceeding its natural sequestration. This large and accelerating transfer of carbon to the atmosphere and upper ocean may lead to global environmental changes that would adversely impact our quality of life. This diagram traces the flow of carbon through the subsurface, the biosphere, and the human energy system. The energy conversions that lead to the sources of anthropogenic carbon dioxide are illustrated. In conjunction with the global exergy diagram above, alternate energy pathways that either do not require fossil carbon or store it in reservoirs other than the atmosphere and ocean surface can be examined in the context of current exergy use and carbon flow. Source: Wes Hermann and A. J. Simon, Global Climate and Energy Project at Stanford University (http://gcep.stanford.edu), © 2007.

without more explanation than I can provide here, what it shows are all possible energy sources on planet earth, including fossil fuels—which are only a tiny fraction of our possible energy sources. We have many other sources of energy.

These visualizations of our energy sources and uses allow us to imagine that if we electrify everything, we will need far less energy to begin with. The Sankey diagrams give us the opportunity to clearly see a carbon-free future.

# E

## DOWN THE RABBIT HOLE: LOOK FOR YOURSELF

Energy Information Administration (EIA), Monthly Energy Review, https://www.eia.gov/totalenergy/data/monthly/

EIA, by Sector Energy Use, https://www.eia.gov/totalenergy/data/annual/

EIA, Manufacturing Energy Consumption Survey, https://www.eia.gov/consumption/manufacturing/

EIA, Residential Energy Consumption Survey, https://www.eia.gov/consumption/residential/about.php

EIA, Commercial Business Energy Consumption Survey, https://www.eia.gov/consumption/commercial/about.php

Environmental Protection Agency (EPA), Greenhouse Gas Inventory Data Explorer, https://cfpub.epa.gov/ghgdata/inventoryexplorer/

Federal Energy Management Program (FEMP), https://energy.gov/eere/femp/federal-energy-management-program

Material Flow Analysis Reporter, http://www.materialflows.net/visualisation-centre/raw-material-profiles/

Oak Ridge National Laboratory (ORNL), National Household Transit Survey, http://nhts.ornl.gov/

ORNL, Transportation Energy Data Book, http://cta.ornl.gov/data/index.shtml

US Consumer Expenditure Surveys, https://www.bls.gov/cex/

US Current Employment Statistics, https://www.bls.gov/ces/

US Unemployment Data, https://www.bls.gov/cps/tables.htm

# ACKNOWLEDGMENTS

Many great people have helped with or influenced this book. Thanks to you all for inspiration, feedback, and help: Martha Amram, David Benzler, Arjun Bhargava, Clayton Boyd, Dane Boysen, Stewart Brand, Steve Chu, Simon Clark, Hans von Clemm, Lisa Cunningham, Nick Dragotta, Mark Duda, Drew Endy, Jacob Friedman, Todd Georgopapadakos, Jennifer Gerbi, Tucker Gilman, Pamela Griffith, Arwen O'Reilly Griffith, Ross Griffith, Selena Griffith, Bronte Griffith, Huxley Griffith, Paul Hawken, Joanne Huang, Christina Isobel, Nathanael Johnson, Alex Kaufman, Kevin Kelly, Jonathan Koomey, Alex Laskey, Emily Leslie, Patti Lord, Pete Lynn, David JC MacKay (in memory), Leila Madrone, Deborah Marshall (in memory), Yaron Milgrom-Elcott, Tim Newell, Tim O'Reilly, Jen Pahlka, Dan Recht, Kirk von Rohr, Vince Romanin, Gwen Rose, Joel Rosenberg, Jason Rugolo, Caroline Spears, Nat F. C. Torkington, Ron Trauner, George Warner, Jason Wexler, Eric Wilhelm, Seth Zuckerman, Adam Zurofsky.

But very special thanks to Sam Calisch and Laura Fraser. Sam was more like a coauthor and waded through the data trenches, endless spreadsheets, and visualization scripts. Similarly, Laura taught both Sam and me how to hit writing deadlines while helping us infuse good grammar, good spelling, and a bit of fun and love into a topic that could otherwise be a tad dry. Also, thanks to Keith Pasko, Jim McBride, and Pushan Panda for help with data, numbers, Latex, graphics, databases and web scrapes, and with understanding this big challenge which is energy. Lastly, thanks to Neil Gershenfeld, my old professor at MIT, for his introduction to MIT Press, and to the staff there for taking this book on, for being a pleasure to work with, and for being patient (my new favorite word is "stet"). Beth Clevenger, a deft editor, made the book much better. Will Myers and Virginia Crossman are incredibly eagle-eyed copy editors. Hearty appreciation to Anthony Zannino, Sean Reilly, designer Marge Encomienda, Janet Rossi, publicist Heather Goss, and everyone else who helped get this manuscript over the line.

# NOTES

**CHAPTER 1**

1. World Resources Institute, "World Greenhouse Gas Emissions: 2016," February 2020, https://www.wri.org/resources/data-visualizations/world-greenhouse-gas-emissions -2016.

**CHAPTER 2**

1. Pew Research Center, *Americans, Politics, and Science Issues*, July 1, 2015, 89, https:// www.pewresearch.org/internet/wp-content/uploads/sites/9/2015/07/2015-07-01 _science-and-politics_FINAL-1.pdf.

2. Paige Hanley, "Pope Demands Action for Failing Fight against Climate Change," Catholic News Service, December 4, 2019, https://www.catholicnews.com/pope -demands-action-for-failing-fight-against-climate-change/.

3. Jim Tolbert, "Republicans Came to the Table on Climate This Year," *The Hill*, December 30, 2019, https://thehill.com/blogs/congress-blog/energy-environment /476210-republicans-came-to-the-table-on-climate-this-year.

4. Andrew Rafferty and Ellen Rolfes, "How Young Conservatives Hope to Make Climate a GOP Issue," Newsy, September 4, 2019, https://www.newsy.com/stories /how-conservatives-are-trying-to-make-climate-a-gop-issue/.

5. Pew Research Center, *US Public Views on Climate and Energy*, November 25, 2019, 2, https://www.pewresearch.org/science/wp-content/uploads/sites/16/2019/11 /Climate-Energy-REPORT-11-22-19-FINAL-for-web-1.pdf.

6. Jeff McMahon, "Former UN Climate Chief Calls for Civil Disobedience," *Forbes*, February 24, 2020, https://www.forbes.com/sites/jeffmcmahon/2020/02/24/former -un-climate-chief-calls-for-civil-disobedience/.

7. Cara Buckley, "Jane Fonda at 81, Proudly Protesting and Going to Jail," *New York Times*, November 3, 2019, https://www.nytimes.com/2019/11/03/arts/television/04 jane-fonda-arrest-protest.html.

8. Frederica Perera, "Pollution from Fossil-Fuel Combustion is the Leading Environmental Threat to Global Pediatric Health and Equity: Solutions Exist," *International Journal of Environmental Research and Public Health* 15, no. 1 (January 2018): 16, https://www.ncbi.nlm.nih.gov/pmc/articles/PMC5800116/.

9. Paris Agreement, Chapter XXVII, 7.d., United Nations Treaty Collection, December, 12, 2015, https://unfccc.int/sites/default/files/english_paris_agreement.pdf.

10. Intergovernmental Panel on Climate Change, *Global Warming of 1.5°C*, retrieved October 7, 2018, https://www.ipcc.ch/sr15/.

11. Intergovernmental Panel on Climate Change, *Global Warming of 1.5°C*.

12. Kurt Zenz House et al., "Economic and Energetic Analysis of Capturing $CO_2$ from Ambient Air," *Proceedings of the National Academy of Sciences* 108, no. 51 (December 20, 2011), https://doi.org/10.1073/pnas.1012253108.

13. Timothy N. Lenton, "Climate Tipping Points—Too Risky to Bet Against," *Nature*, November 27, 2019, https://www.nature.com/articles/d41586-019-03595-0.

14. Michaela D. King et al., "Dynamic Ice Loss from the Greenland Ice Sheet Driven by Sustained Glacier Retreat," *Communications Earth and Environment* 1, no. 1 (August 2020), https://doi.org/10.1038/s43247-020-0001-2.

15. Zeke Hausfather, "UNEP: 1.5C Climate Target 'Slipping out of Reach,'" Carbon Brief, November 26, 2019, https://www.carbonbrief.org/unep-1-5c-climate-target -slipping-out-of-reach.

16. Robbie Andrew, "It's Getting Harder and Harder to Limit Ourselves to 2°C," Desdemona Despair, April 23, 2020, https://desdemonadespair.net/2019/08/its -getting-harder-and-harder-to-limit-global-warming-to-2c-it-is-partly-this-hope-in-future -technologies-that-delays-action.html.

17. Johan Rockström et al., "A Roadmap for Rapid Decarbonization," *Science* 355, no. 6,331 (March 24, 2017): 1,269.

18. Dan Tong et al., "Committed Emissions from Existing Energy Infrastructure Jeopardize 1.5°C Climate Target," *Nature* 572, no. 7,769 (August 2019): 373–377, https://www.nature.com/articles/s41586-019-1364-3.

19. "In 2018, 66% of New Electricity Generation Capacity Was Renewable, Price of Batteries Dropped 35%," *SDG Knowledge Hub* (blog), "International Institute for Sustainable Development, April 9, 2019, https://sdg.iisd.org/news/in-2018-66-of-new -electricity-generation-capacity-was-ren.

20. National Association of Home Builders and Bank of America Home Equity, *Study of Life Expectancy of Home Components*, February 2007, https://www.interstatebrick .com/sites/default/files/library/nahb20study20of20life20expectancy20of20home20 components.pdf; "By the Numbers: How Long Will Your Appliances Last? It Depends," *Consumer Reports*, March 21, 2009, https://www.consumerreports.org/cro

/news/2009/03/by-the-numbers-how-long-will-your-appliances-last-it-depends
/index.htm.

## CHAPTER 3

1. Dayton Duncan and Ken Burns, *The National Parks: America's Best Idea, An Illustrated History* (New York: Alfred A. Knopf, 2009).

2. National Park Service, "100th Anniversary of President Theodore Roosevelt and Naturalist John Muir's Visit at Yosemite National Park," news release, May 13, 2003, quoted in National Park Service, "John Muir," https://www.nps.gov/jomu/learn /historyculture/people.htm.

3. Michelle Mock, "The Electric Home and Farm Authority, 'Model T Appliances,' and the Modernization of the Home Kitchen in the South," *The Journal of Southern History* 80, no. 1 (February 2014): 73–108, https://www.jstor.org/stable/23796844.

4. US Department of Energy, "FY 2020 Budget Request Fact Sheet," March 11, 2019, https://www.energy.gov/articles/department-energy-fy-2020-budget-request-fact -sheet.

5. Centers for Disease Control and Prevention, "History of the Surgeon General's Reports on Smoking and Health," November 15, 2019, https://www.cdc.gov /tobacco/data_statistics/sgr/history/index.htm.

6. Theodore R. Holford et al., "Tobacco Control and the Reduction in Smoking-Related Premature Deaths in the United States, 1964–2012," *JAMA* 311, no. 2 (2014): 164–171, https://doi.org/10.1001/jama.2013.285112.

7. World Health Organization, "Health Benefits Far Outweigh the Costs of Meeting Climate Change Goals," news release, December 5, 2018, https://www.who.int /news/item/05-12-2018-health-benefits-far-outweigh-the-costs-of-meeting-climate -change-goals.

8. US Environmental Protection Agency, "Climate Impacts on Human Health," January 19, 2017, https://19january2017snapshot.epa.gov/climate-impacts/climate -impacts-human-health_.html.

9. "Montreal Protocol," Wikipedia, https://en.wikipedia.org/wiki/Montreal_Protocol.

10. J. Maxwell and F. Briscoe, "There's Money in the Air: The CFC Ban and DuPont's Regulatory Strategy," *Business Strategy and the Environment* 6, no. 5 (January 1997): 276–286, https://doi.org/10.1002/(SICI)1099-0836(199711)6:5<276::AID-BSE123> 3.0.CO;2-A.

11. Chandra Bhushan, "A Monopoly Like None Other," Down to Earth, April 20, 2016, https://www.downtoearth.org.in/blog/climate-change/a-monopoly-like-none -other-53610.

12. Sharon Lerner, "How a DuPont Spinoff Lobbied the EPA to Stave Off the Use of Environmentally-Friendly Coolants," *The Intercept*, August 25, 2018, https:// theintercept.com/2018/08/25/chemours-epa-coolant-refrigerant-dupont/.

## CHAPTER 4

1. US Congress Joint Committee on Atomic Energy, *Understanding the "National Energy Dilemma"* (Washington: The Center for Strategic and International Studies, 1973).

2. A. L. Austin and S. D. Winter, *US Energy Flow Charts for 1950, 1960, 1970, 1980 and 1990* (Livermore, CA: Lawrence Livermore National Laboratory, 1973).

3. US Energy Information Administration, Manufacturing Energy Consumption Survey (MECS) 2014, September 6, 2018, https://www.eia.gov/consumption/manu facturing/data/2018/.

4. US Energy Information Administration, Residential Energy Consumption Survey (RECS) 2015, https://www.eia.gov/consumption/residential/.https://www.eia.gov/con sumption/residential/.

5. US Energy Information Administration, Commercial Buildings Energy Consumption Survey (CBECS) 2012, https://www.eia.gov/consumption/commercial/data/2012/.

6. US Department of Transportation, National Household Travel Survey, 2017, https://nhts.ornl.gov/.

7. Office of NEPA Policy and Compliance, US Department of Energy, *An Open-Source Tool for Visualizing Energy Data to Identify Opportunities, Inform Policy, and Increase Energy Literacy*, Advanced Research Projects Agency (DOE), n.d., Project Grant DEAR0000853, https://www.energy.gov/nepa/downloads/cx-016689-open-source-tool-visualizing -energy-data-identify-opportunities-inform.

8. Eric Masanet et al., "Recalibrating Global Data Center Energy-Use Estimates," *Science* 367, no. 6,481 (February 28, 2020): 984–986.

## CHAPTER 5

1. US Environmental Protection Agency, "Evolution of the Clean Air Act," https://www.epa.gov/clean-air-act-overview/evolution-clean-air-act.

2. An Act to Amend the Federal Water Pollution Control Act, Pub. L. No. 92-500, October 18, 1972.

3. Edward Cowan, "President Urges 65° as Top Heat in Homes to Ease Energy Crisis," *New York Times*, January 22, 1977, https://www.nytimes.com/1977/01/22 /archives/president-urges-65-as-top-heat-in-homes-to-ease-energy-crisis-cites.html; "Transcript of Nixon's Speech on Energy Situation," *New York Times*, January 20, 1974: 36, https://timesmachine.nytimes.com/timesmachine/1974/01/20/93255285 .html?pageNumber=36.

## CHAPTER 6

1. US Department of Energy, *2016 Billion-Ton Report: Advancing Domestic Resources for a Thriving Bioeconomy*, Volume I, July 2016, https://www.energy.gov/sites/prod /files/2016/12/f34/2016_billion_ton_report_12.2.16_0.pdf.

2. Paul E. Brockway et al., "Estimation of Global Final-Stage Energy-Return-on-Investment for Fossil Fuels with Comparison to Renewable Energy Sources," *Nature Energy* 4 (July 2019): 612–621, https://www.nature.com/articles/s41560-019-0425-z.

## CHAPTER 7

1. David J. C. MacKay, *Sustainable Energy—Without the Hot Air* (Cambridge, UK: UIT Cambridge, 2009), 33.

2. US Department of Agriculture, "Feedgrains Sector at a Glance," October 23, 2020, https://www.ers.usda.gov/topics/crops/corn-and-other-feedgrains/feedgrains-sector -at-a-glance/.

3. S. De Stercke, *Dynamics of Energy Systems: A Useful Perspective*, IIASA Interim Report No. IR-14-013 (Laxenburg, Austria: International Institute for Applied Systems Analysis, 2014).

4. Steve Hanley, "New Mark Z. Jacobson Study Draws A Roadmap To 100% Renewable Energy," CleanTechnica, February 8, 2018, https://cleantechnica.com /2018/02/08/new-jacobson-study-draws-road-map-100-renewable-energy/.

5. Mark Z. Jacobson et al., "Low-Cost Solution to the Grid Reliability Problem with 100% Penetration of Intermittent Wind, Water, and Solar for All Purposes," *Proceedings of the National Academy of Sciences* 112, no. 49 (December 8, 2015): 15,060–15,065, https://www.pnas.org/content/112/49/15060.

6. Christopher T. M. Clack et al., "Evaluation of a Proposal for Reliable Low-Cost Grid Power with 100% Wind, Water, and Solar," *Proceedings of the National Academy of Sciences* 114, no. 26 (June 27, 2017): 6,722–6,727, https://www.pnas.org /content/114/26/6722.

7. Mark Z. Jacobson et al., "The United States Can Keep the Grid Stable at Low Cost with 100% Clean, Renewable Energy in All Sectors Despite Inaccurate Claims," *Proceedings of the National Academy of Sciences* 114, no. 26 (June 27, 2017): E5,021–E5,023, https://www.pnas.org/content/114/26/E5021.

8. *Response to Jacobson et al. (June 2017)*, Dr. Staffan Qvist, https://www.vibrant cleanenergy.com/wp-content/uploads/2017/06/ReplyResponse.pdf.

9. National Renewable Energy Laboratory, *Renewable Electricity Futures Study Volume 1: Exploration of High-Penetration Renewable Electricity Futures*, US Department of Energy, 2012, https://www.nrel.gov/docs/fy12osti/52409-1.pdf.

10. Steve Fetter, "How Long Will the World's Uranium Supplies Last?," *Scientific American*, January 26, 2009, https://www.scientificamerican.com/article/how-long -will-global-uranium-deposits-last/.

11. Thomas Wellock, "'Too Cheap to Meter': A History of the Phrase," US Nuclear Regulatory Commission Blog, June 3, 2016, https://public-blog.nrc-gateway.gov/2016 /06/03/too-cheap-to-meter-a-history-of-the-phrase/.

12. Mark Diesendorf, "Dispelling the Nuclear Baseload Myth: Nothing Renewables Can't Do Better," Energy Post, March 23, 2016, https://energypost.eu/dispelling -nuclear-baseload-myth-nothing-renewables-cant-better/.

13. "Watts Bar Nuclear Plant," Wikipedia, https://en.wikipedia.org/wiki/Watts_Bar _Nuclear_Plant.

14. US Energy Information Agency, "Nuclear Explained: US Nuclear Industry," April 15, 2020, https://www.eia.gov/energyexplained/nuclear/us-nuclear-industry.php.

15. Office of Energy Efficiency and Renewable Energy, "Solar Energy Technologies Office Fiscal Year 2019 Funding Program," US Department of Energy, 2019, https://eere-exchange.energy.gov/FileContent.aspx?FileID=d2f56f78-decb-4cc1-9a88-0f6241708508.

## CHAPTER 8

1. Kevin Ridder, "The Problem with Monopoly Utilities," *The Appalachian Voice*, October 17, 2018, https://appvoices.org/2018/10/17/the-problem-with-monopoly-utilities/.

2. "Hills Hoist," Wikipedia, https://en.wikipedia.org/wiki/Hills_Hoist.

3. US Energy Information Administration, "Figure 2.1: Energy Consumption by Sector," *Monthly Energy Review*, February 2021, https://www.eia.gov/totalenergy/data/monthly/pdf/sec2_2.pdf.

4. "Underground Natural Gas Storage," Energy Infrastructure, 2020, https://energy infrastructure.org/energy-101/natural-gas-storage.

5. US Energy Information Administration, "Table 6.3: Coal Stocks by Sector," *Monthly Energy Review*, February 2021, https://www.eia.gov/totalenergy/data/monthly/pdf/sec6_5.pdf; "Coal Stockpiles at US Coal Power Plants Were at Their Lowest Point in Over a Decade," *Today in Energy* (blog), US Energy Information Administration, May 27, 2019, https://www.eia.gov/todayinenergy/detail.php?id=39512.

6. Noah Kittner, Felix Lill, and Daniel M. Kammen, "Energy Storage Deployment and Innovation for the Clean Energy Transition," *Nature Energy* 2, no. 17,125 (July 31, 2017), https://escholarship.org/content/qt62d4075g/qt62d4075g_noSplash_c77f50 aad68d476a1432e17518430dac.pdf; Logan Goldie-Scot, "A Behind the Scenes Take on Lithium-ion Battery Prices," BloombergNEF, March 5, 2019, https://about.bnef .com/blog/behind-scenes-take-lithium-ion-battery-prices/.

7. MacKay, *Sustainable Energy*, 153.

8. Office of Energy Efficiency and Renewable Energy, "Energy Analysis, Data, and Reports: Manufacturing Energy Bandwidth Studies," US Department of Energy, 2013, https://www.energy.gov/eere/amo/energy-analysis-data-and-reports.

9. "Atlas of 100% Renewable Energy," Wärtsilä, 2020, https://www.wartsila.com /energy/towards-100-renewable-energy/atlas-of-100-percent-renewable-energy#/.

10. US Energy Information Administration, "Table 10.1: Renewable Energy Production and Consumption by Source," *Monthly Energy Review*, February 2021, https://www.eia.gov/totalenergy/data/monthly/pdf/sec10_3.pdf.

## CHAPTER 10

1. *Lazard's Levelized Cost of Energy Analysis,* Version 13, Lazard, November 7, 2019, https://www.lazard.com/perspective/lcoe2019.

2. Office of Energy Efficiency and Renewable Energy, "Soft Costs," US Department of Energy, 2020, https://www.energy.gov/eere/solar/soft-costs.

3. Edward Rubin et al., "A Review of Learning Rates for Electricity Supply Technologies," *Energy Policy* 86 (November 2015): 198–218, https://www.cmu.edu/epp/iecm /rubin/PDF%20files/2015/A%20review%20of%20learning%20rates%20for%20 electricity%20supply%20technologies.pdf.

4. T. P. Wright, "Factors Affecting the Cost of Airplanes," *Journal of Aeronautical Sciences* 3, no. 4 (February 1936), https://arc.aiaa.org/doi/10.2514/8.155.

5. Gordon E. Moore, "Cramming More Components onto Integrated Circuits," *Electronics*, April 19, 1965, 114–117.

6. Béla Nagy et al., "Statistical Basis for Predicting Technological Progress," *PLOS One* 8, no. 2 (February 23, 2013): e52,669, https://journals.plos.org/plosone/article ?id=10.1371/journal.pone.0052669.

7. Rubin, "A Review of Learning Rates," 198–218.

8. "Sunny Uplands," *The Economist*, November 21, 2012, https://www.economist.com /news/2012/11/21/sunny-uplands.

9. Nancy M. Haegel et al., "Terawatt-Scale Photovoltaics: Trajectories and Challenges," *Science* 356, no. 6,334 (April 14, 2017): 141–143, https://science.sciencemag.org /content/356/6334/141.summary.

10. International Renewable Energy Agency, "Renewable Energy Now Accounts for a Third of Global Power Capacity," news release, April 2, 2019, https://www .irena.org/newsroom/pressreleases/2019/Apr/Renewable-Energy-Now-Accounts-for -a-Third-of-Global-Power-Capacity.

11. The exact number depends on how world population grows, and what quality of life is enjoyed by what percentage of humans. De Stercke, *Dynamics of Energy Systems*.

## CHAPTER 11

1. Saul Griffith and Sam Calisch, "No Place Like Home: Fighting Climate Change (and Saving Money) by Electrifying America's Households," Rewiring America, October 2020, https://www.rewiringamerica.org/household-savings-report.

2. US Bureau of Labor Statistics, "Consumer Expenditure Surveys: State-Level Expenditure Tables by Income," 2020, https://www.bls.gov/cex/csxresearchtables.htm.

3. US Energy Information Administration, "State Energy Data System (SEDS): 1960–2018 (complete)," 2020, https://www.eia.gov/state/seds/seds-data-complete.php.

4. Office of Energy Efficiency and Renewable Energy, "Find and Compare Cars: 2020 Honda Civic 4Dr," US Department of Energy, https://www.fueleconomy.gov /feg/noframes/42150.shtml.

5. Office of Energy Efficiency and Renewable Energy, "Find and Compare Cars: 2019 BMW 540i," US Department of Energy, https://www.fueleconomy.gov/feg /noframes/40477.shtml.

6. Office of Energy Efficiency and Renewable Energy, "Find and Compare Cars: 2019 Chevrolet Silverado LD C15 2WD," US Department of Energy, https://www .fueleconomy.gov/feg/noframes/40258.shtml.

7. National Renewable Energy Laboratory, "Typical Meteorological Year (TMY)," National Solar Radiation Database, https://nsrdb.nrel.gov/about/tmy.html.

8. Sanden Water Heater, "Sanden SANCO2: Heat Pump Water Heater Technical Information," Sanden Water Heater, October 2017.

9. Office of Energy Efficiency & Renewable Energy (EERE), "Commercial and Residential Hourly Load Profiles for all TMY3 Locations in the United States," US Department of Energy, last updated July 2, 2013, https://openei.org/doe-opendata /dataset/commercial-and-residential-hourly-load-profiles-for-all-tmy3-locations-in -the-united-states.

10. Heather Lammers, "News Release: NREL Raises Rooftop Photovoltaic Technical Potential Estimate," National Renewable Energy Laboratory, March 24, 2016, https://www.nrel.gov/news/press/2016/24662.html.

## CHAPTER 12

1. "Home Owners' Loan Act (1933)," The Living New Deal, 2020, https://livingnew deal.org/glossary/home-owners-loan-act-1933/.

2. Mock, "The Electric Home and Farm Authority."

## CHAPTER 13

1. James McKellar, "Oil and Gas Financing: 'How It Works'" (presentation, 32nd Annual Ernest E. Smith Oil, Gas, & Mineral Law Institute, Houston, TX, March 31, 2006).

2. R. Allen Myles et al., "Warming Caused by Cumulative Carbon Emissions towards the Trillionth Tonne," *Nature* 458, no. 7,242 (May 2009): 1,163–1,166.

3. Office of Energy Efficiency & Renewable Energy (EERE), *Manufacturing Energy Bandwidth Studies* (2014 MECS), Energy Analysis, Data and Reports, US Department of Energy, https://www.energy.gov/eere/amo/energy-analysis-data-and-reports.

4. J. F. Mercure et al., "Macroeconomic Impact of Stranded Fossil Fuel Assets," *Nature Climate Change* 8 (2018): 588–593, https://doi.org/10.1038/s41558-018-0182-1.

5. Richard Knight, "Sanctions, Disinvestment, and US Corporations in South Africa," in *Sanctioning Apartheid*, ed. Robert E. Edgar (Trenton, NJ: Africa World Books, 1991).

6. "Oil Company Earnings: Reality Over Rhetoric," *Forbes*, May 10, 2011, https:// www.forbes.com/2011/05/10/oil-company-earnings.html.

## CHAPTER 14

1. Jon Henley and Elisabeth Ulven, "Norway and the A-ha Moment that Made Electric Cars the Answer," *The Guardian*, April 19, 2020, https://www.theguardian .com/environment/2020/apr/19/norway-and-the-a-ha-moment-that-made-electric -cars-the-answer.

2. California Energy Commission, "2019 Building Energy Efficiency Standards," 2020, https://www.energy.ca.gov/programs-and-topics/programs/building-energy-efficiency-standards/2019-building-energy-efficiency.

3. San Francisco Planning Department, *Zoning Administrator Bulletin No. 11: Better Roofs Ordinance*, 2019, https://sfplanning.org/sites/default/files/documents/publications/ZAB_11_Better_Roofs.pdf.

4. Susie Cagle, "Berkeley Became First US City to Ban Natural Gas. Here's What That May Mean for the Future," *The Guardian*, July 23, 2019, https://www.theguardian.com/environment/2019/jul/23/berkeley-natural-gas-ban-environment.

5. Chris D'Angelo, "The Gas Industry's Bid to Kill A Town's Fossil Fuel Ban," *Huffington Post*, December 16, 2019, https://www.huffpost.com/entry/massachusetts-natural-gas-ban_n_5de93ae2e4b0913e6f8ce07d.

6. "Net Metering," Solar Energy Industries Association, 2020, https://www.seia.org/initiatives/net-metering.

7. California Public Utilities Commission, "What Are TOU Rates?," 2020, https://www.cpuc.ca.gov/general.aspx?id=12194.

8. Legal Pathways to Deep Decarbonization, https://lpdd.org.

## CHAPTER 15

1. Richard Scarry, *What Do People Do All Day?* (New York: Golden Books, 1968).

2. "Fact Sheets," National Association of Convenience Stores, 2020, https://www.convenience.org/Research/FactSheets.

3. Saul Griffith and Sam Calisch, "Mobilizing for a Zero-Carbon America: Jobs, Jobs, and More Jobs," Rewiring America, July 2020, https://www.rewiringamerica.org/jobs-report.

4. Arthur Herman, *Freedom's Forge: How American Business Produced Victory in World War II* (New York: Random House, 2012); US War Production Board, *Wartime Production Achievements and the Reconversion Outlook: Report of the Chairman*, October 9, 1945, https://catalog.hathitrust.org/Record/001313077.

## CHAPTER 16

1. "We Shall Fight on the Beaches," International Churchill Society, 2020, https://winstonchurchill.org/resources/speeches/1940-the-finest-hour/we-shall-fight-on-the-beaches/.

2. *Journal of the House of Representatives of the United States,* 77th Congress, Second Session, January 5, 1942 (Washington, DC: US Government Printing Office, 1942), 6; emphasis mine.

3. William M. Franklin and William Gerber, eds., *Foreign Relations of the United States: Diplomatic Papers, The Conferences at Cairo and Tehran, 1943,* President's Log at Tehran entry on Tuesday, November 30, 8:30 p.m. (Washington DC: US Government Printing Office, 1961), 469, https://history.state.gov/historicaldocuments/frus1943CairoTehran/d353.

**CHAPTER 17**

1. Nicholas Rees and Richard Fuller, *The Toxic Truth: Children's Exposure to Lead Pollution Undermines a Generation of Future Potential*, UNICEF and Pure Earth, 2020, https://www.unicef.org/media/73246/file/The-toxic-truth-children's-exposure-to -lead-pollution-2020.pdf.

2. Sérgio Faias, Jorge Sousa, Luís Xavier, and Pedro Ferreira, "Energy Consumption and $CO_2$ Emissions Evaluation for Electric and Internal Combustion Vehicles Using a LCA Approach," *Renewable Energies and Power Quality Journal* 1, no. 9 (May 2011): 1382–1388, http://www.icrepq.com/icrepq'11/660-faias.pdf.

3. Office of Energy Efficiency and Renewable Energy, "Energy Analysis, Data and Reports," US Department of Energy, 2020, https://www.energy.gov/eere/amo/energy -analysis-data-and-reports.

4. Stephen Nellis, "Apple Buys First-Ever Carbon-Free Aluminum from Alcoa-Rio Tinto Venture," Reuters, December 5, 2019, https://www.reuters.com/article/us -apple-aluminum/apple-buys-first-ever-carbon-free-aluminum-from-alcoa-rio-tinto -venture-idUSKBN1Y91RQ.

5. Center for International Environmental Law, *Plastic & Climate: The Hidden Costs of a Plastic Planet*, May 2019, https://www.ciel.org/wp-content/uploads/2019/05 /Plastic-and-Climate-FINAL-2019.pdf.

6. CIEL, *Plastic & Climate*.


**APPENDIX A**

1. Ben Blatt, "*Where's Waldo*'s Elusive Hero Didn't Just Get Sneakier. He Got Smaller," *Slate*, March 7, 2017, https://slate.com/culture/2017/03/where-s-waldo-didn-t-just -get-harder-to-find-he-got-80-percent-smaller.html.

2. House, "Economic and Energetic Analysis."

3. World Resources Institute, "World Greenhouse Gas Emissions: 2016."

4. Mustapha Harb et al., "Projecting Travelers into a World of Self-Driving Vehicles: Estimating Travel Behavior Implications via a Naturalistic Experiment," *Transportation* 45, no. 6 (November 2018): 1,671–1,685.


**APPENDIX B**

1. Joni Mitchell, "Big Yellow Taxi," *Ladies of the Canyon* (1970).


**APPENDIX C**

1. Jack Pales and Charles Keeling, "The Concentration of Atmospheric Carbon Dioxide in Hawaii," *Journal of Geophysical Research* 70, no. 24, 1965.

2. Pieter Tans and Ralph Keeling, "Mauna Loa $CO_2$ Monthly Mean Concentration," Wikimedia Commons, January 6, 2019, https://commons.wikimedia.org/w/index .php?curid=40636957.

3. Syukuro Manabe and Richard T. Wetherald, "Thermal Equilibrium of the Atmosphere with a Given Distribution of Relative Humidity," *Journal of the Atmospheric Sciences* 24, no. 3, 1967.

4. William W. L. Cheung et al., "Large-Scale Redistribution of Maximum Fisheries Catch Potential in the Global Ocean under Climate Change," *Global Change Biology* 16, no. 1, January 2010, https://onlinelibrary.wiley.com/doi/abs/10.1111/j.1365 -2486.2009.01995.x; Cynthia Rosenzweig et al., "Assessing Agricultural Risks of Climate Change in the 21st Century in a Global Gridded Crop Model Intercomparison," *Proceedings of the National Academy of Sciences* 111, no. 9 (March 4, 2014), https://www.pnas.org/content/111/9/3268.

5. Daniel Scott and Stefan Gössling, *Tourism and Climate Mitigation: Embracing the Paris Agreement*, Brussels: European Travel Commission, March 2018, https://etc -corporate.org/uploads/2018/03/ETC-Climate-Change-Report_FINAL.pdf.

6. Stephane Hallegatte et al., *Shock Waves: Managing the Impacts of Climate Change on Poverty* (Washington, DC: World Bank, 2016), https://openknowledge.worldbank .org/handle/10986/22787.

7. Calum T. M. Nicholson, "Climate Change and the Politics of Casual Reasoning: The Case of Climate Change and Migration," *The Geographical Journal* 180, no. 2 (June 2014), https://rgs-ibg.onlinelibrary.wiley.com/doi/abs/10.1111/geoj.12062.

8. Solomon Hsiang et al., "Estimating Economic Damage from Climate Change in the United States," *Science* 356, no. 6,345 (June 30, 2017): 1,362–1,369, https:// science.sciencemag.org/content/356/6345/1362.

9. Solomon Hsiang and Marshall Burke, "Climate, Conflict, and Social Stability: What Does the Evidence Say?," *Climatic Change* 123 (2014): 39–55, https://link .springer.com/article/10.1007/s10584-013-0868-3.

10. Marko Tainio, "Future Climate and Adverse Health Effects Caused by Fine Particulate Matter Air Pollution: Case Study for Poland," *Regional Environmental Change* 13 (2013): 705–715, https://link.springer.com/article/10.1007/s10113-012-0366-6.

11. Zhoupeng Ren et al., "Predicting Malaria Vector Distribution under Climate Change Scenarios in China: Challenges for Malaria Elimination," *Scientific Reports* 6, no. 20,604 (February 12, 2016), https://www.ncbi.nlm.nih.gov/pmc/articles /PMC4751525/.

12. Tord Kjellstrom, R. Sari Kovats, Simon J. Lloyd, Tom Holt, and Richard S. J. Tol, "The Direct Impact of Climate Change on Regional Labor Productivity," *Archives of Environmental and Occupational Health* 64, no. 4 (Winter 2009): 217–227, doi: 10.1080/19338240903352776.

13. Ove Hoegh-Guldberg et al., "Impacts of 1.5°C Global Warming on Natural and Human Systems," in *Global Warming of 1.5°C*, eds. Valérie Masson-Delmotte et al., Intergovernmental Panel on Climate Change, 2019, https://www.ipcc.ch/site/assets /uploads/sites/2/2019/06/SR15_Chapter3_Low_Res.pdf.

14. R. Allen Myles et al., "Warming Caused by Cumulative Carbon Emissions towards the Trillionth Tonne," *Nature* 458, no. 7,242 (May 2009): 1,163–1,166.

**APPENDIX D**

1. Sandra Rendgen, *The Minard System: The Complete Statistical Graphics of Charles-Joseph Minard* (Princeton, NJ: Princeton Architectural Press, 2018).

2. US Bureau of Labor Statistics, "Consumer Expenditures—2019," news release, September 9, 2020, https://www.bls.gov/news.release/cesan.nr0.htm.

3. Lawrence Livermore National Laboratory, "How to Read an LLNL Energy Flow Chart (Sankey Diagram)," YouTube, April 19, 2016, https://www.youtube.com/watch ?v=G6dlvECRfcI.

# INDEX

Page numbers followed by "f" and "t" refer to figures and tables, respectively.